54
THE SATURDAY EVENING POST
May 4, 1912
$84.30 for this new
including the "M 10" series of Columbia Double-Disc Records and Record Album
Until July 1, delivered ON APPROVAL by 7500 dealers, payment terms $7 a month—no interest, no extras—and subject to a guarantee of full satisfaction.
Or $75 flat for the instrument alone. No other upright instrument has ever been offered for less than $100, with NO records.
THIS new Grafonola "Princess" is a beautiful mahogany table, ready at all times to serve all the purposes of such a table, worthy of its place in any part of any room in the house, yet likewise ready to provide music of any class you happen to desire, and capable of all the quality of tone of any instrument by which you may happen to judge it.
This newest musical instrument, with the "M 10" outfit of records—a total of 24 different selections, sufficient for an entire evening's delightful entertainment, together with an album to contain them—costs you just $84.30, and until July 1st, may be paid for at the rate of $7 a month, and will be delivered on approval by your dealer. Besides the "M 10" series of records, do not miss the opportunity of hearing this most versatile of all musical instruments render selections by Lillian Nordica, Olive Fremstad, Mary Garden, Emmy Destinn, Alice Nielsen, Zenatello, David Bispham, Maria Gay and any of the other great artists whose voices appeal to you.
Go to Your Nearest Dealer
The nearest one of 7500 dealers will place this complete outfit in your home on free trial, giving you an opportunity to combine an evening's entertainment with a thorough demonstration, and guaranteeing the outfit to be exactly as advertised or money refunded.
If by any chance there is no dealer in your neighborhood who is prepared to supply you, drop us a line and we will see that your inquiry is attended to.
With lid and tone-shutters closed—a completely concealed musical instrument, serving as a handsome small table.
The tone-shutters are opened or closed by turning this little knob.
Columbia Phonograph Company, General
Box 328, Tribune Building, New York
TORONTO—McKinnon Building.
(Prices in Canada plus duty.)
Dealers Everywhere, but the fast-increasing Columbia demand affords splendid opportunities for MORE DEALERS. Write in for trade particulars. Exclusive selling rights granted where we are not actively represented.
Columbia Hornless Graphophone "Lyric"—$25
Columbia Grafonola "Favorite"—$50
Columbia Grafonola "Nonpareil"—$150
Columbia Grafonola "Regent"—$200
THE SATURDAY EVENING POST
55
Columbia Grafonola
The "Princess"
First announcement of the tone-shutters, replacing the two small doors, and providing more sightly, more effective and more convenient control of tone-volume.
By turning the small knob, it is easy to partly or completely close the tone-shutters at the opening of the sound-chamber, regulating the volume of music without affecting its tonal integrity.
An Exclusive Columbia Feature
There are other Columbia Grafonolas at other prices
(Graphophones from $17.50 up)

Phonographs with Flair

A Century of Style in Sound Reproduction

Timothy C. Fabrizio & George F. Paul
With Photographs by the Authors

Schiffer Publishing Ltd

4880 Lower Valley Road, Atglen, PA 19310 USA

Dedication

Dedicated to Edward Pooley, John Herzog, Morris Keen
and the others who created the look of the phonograph.

You'll want to look for the Orchestra... When your authorized Pickering Dealer is demonstrating the revolutionary new ISOPHASE SPEAKER, please, don't look behind it for the orchestra. Take our word for it – these magnificently realistic sounds are coming from the curved diaphragm itself. "Unbelievable," "extraordinary," "breath-taking," and "window-on-the-studio quality" are some of the comments we've heard from dealers and customers alike. But don't be persuaded by mere words alone. Hear it for yourself.
- Pickering & Co. advertisement, *High Fidelity*, November 1956.

Library of Congress Cataloging-in-Publication Data

Fabrizio, Timothy C.
Phonographs with flair: a century of style in sound reproduction/Timothy C. Fabrizio & George F. Paul.
p. cm.
ISBN 0-7643-1281-2
1. Phonograph--History. I. Paul, George F. II. Title.
TS2301.P3 F2995 2001
621.389'33-dc21
00-011721

Designed by "Sue"
Type set in Souvenir Lt Bt

ISBN: 0-7643-1281-2
Printed in China
1 2 3 4

Published by Schiffer Publishing Ltd.
4880 Lower Valley Road
Atglen, PA 19310
Phone: (610) 593-1777; Fax: (610) 593-2002
E-mail: Schifferbk@aol.com
Please visit our web site catalog at www.schifferbooks.com or write for a free catalog.

We are always looking for authors to write books on new and related subjects. If you have an idea for a book, please contact us at the above address.

This book may be purchased from the publisher.
Please include $3.95 for shipping.

In Europe, Schiffer books are distributed by
Bushwood Books
6 Marksbury Ave.
Kew Gardens
Surrey TW9 4JF England
Phone: 44 (0)20-8392-8585; Fax: 44 (0)20-8392-9876
E-mail: Bushwd@aol.com
Free postage in the UK. Europe: Air mail at cost.
Please try your bookstore first.

Contents

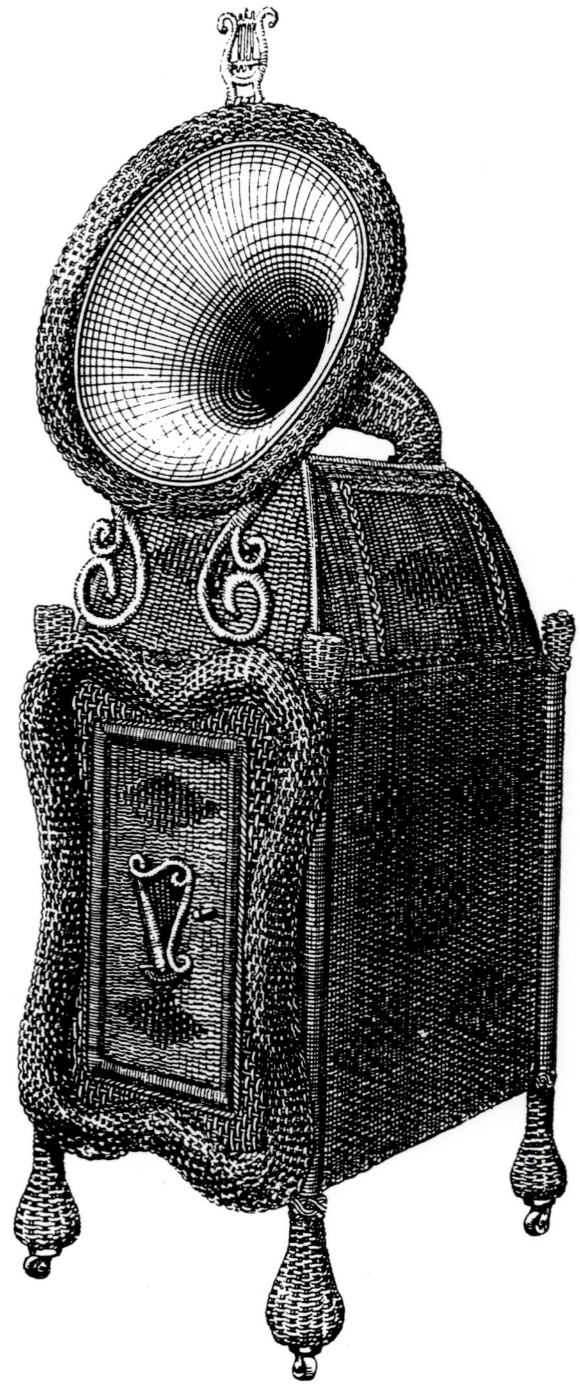

Acknowledgments

The authors would like to thank those collectors and institutions whose items appear anonymously, and under the following attributions: Doug Anderson, John P. Andolina, Jr. "The Early Soundman," Arophon, Chuck Azzalina, Dave Barker, Walt and Ellen Bellm, Omri Ben-Zion, Freddie and Jeannette Blair, Greg Bogantz, Lou Caruso, Gregg Cline, Domenic DiBernardo, Edison National Historic Site, Bob Havalack "Eager Bros. Music," John W. Hess, Fred Guenther, Daniel Gustafson, Ron Haring, Nina and Dave Heitz, Charles Hummel, Garry James, Brady Jeffcoat, Pat and Jamie Jones, Dan and Sandy Krygier, Randy Liebermann, William G. Meyer, Alan H. Mueller, Walter and Carol Myers, Brice D. Paris, John A. Reinicke, Steve and Ellie Saccente, Jasper Sanfilippo, Wilfried Sator, Gus Schild, Norm and Janyne Smith, Gary Stork, Richard Storms, David Werchen, Kurt Wilmarth "Echo-Tone Music," and "Waves." Finally, we are deeply indebted to the National Park Service, Department of the Interior, Edison National Historic Site, and the immense efforts of staff member Jerry Fabris, for providing access to some of the rarest treasures in the Edison legacy.

Photographing books of this sort involves much travel, time and effort. For offering us a place to sleep, feeding us, tolerating us, or otherwise helping us above and beyond the call of duty, we wish to sincerely thank: Jean-Paul and Sylvie Agnard, Julien and Nellie Anton, Greg Bogantz, Dom Coladonato, Scott and Denise Corbett, Ken and Jen Flaherty, Gary and Susan James, Nina and Dave Heitz, Mike and Janene Khanchalian, Philippe and Christine LeRay, Bob and Marilyn Laboda, Rob and Cara Lomas, Charles and Mary McCarn, Tom and Sandy McCarthy, William G. Meyer, Juanita Jones, Pat and Jamie Jones, Sandy and Tessie, Bruce and Linda Peterson, Ray Phillips, Robin and Joan Rolfs, Yves Rouchaleau, Sam and Edna Saccente, Steve and Ellie Saccente, Jasper and Marion Sanfilippo, Sam and Rita Sheena, Norm and Janyne Smith, Marie-Claude Stéger, and David Werchen.

We would also like to extend our gratitude to Bob Baumbach, the unsurpassed W.C. Chin, Stuart Fisher, Tim Gracyk for his able assistance, our indispensable editor Molly Higgins, the ever-enthusiastic Barbara, Jess and Tim Paul, the always-encouraging Joan Paul, the late Dr. John E. Paul, Darren Wallace, and, as always, Ray Wile, archivist supreme.

Very special thanks are due to the following "electric-era" buddies for providing invaluable support and information. Bruce and Charlotte Mager of New York City's "Waves" have repeatedly proved themselves to be not only the finest of friends, but also the foremost of authorities on radios and radio-phonographs. Roger Buttignol took incredible pains to help us in our quest for material to photograph. Greg Bogantz, a veritable font of knowledge on the subject of electric sound reproduction, and a habitué of on-line auctions (giving him a finger on the pulse of the hobby), was immeasurably helpful. Gary Stork, the standard-bearer of the Voice of Music Audio Enthusiasts and publisher of the V-M "Collector's Voice" newsletter, could not have been more energetic and accommodating. Two Rochester, NY, dealers in 45 players, Kurt Wilmarth and Bob Havalack, were very helpful. John Reinicke and others who responded to our notice in the *Antique Radio Classifieds* (even those we were unable to visit in person) are owed a debt that can not be expressed in the few words allowed here.

As is our custom, we would like to honor the wonderful staff at Schiffer Publishing: Peter and Nancy Schiffer, Doug Congdon-Martin and all the other friendly and cooperative folks who work there.

Introduction: A Century of Style, 1877 - 1977

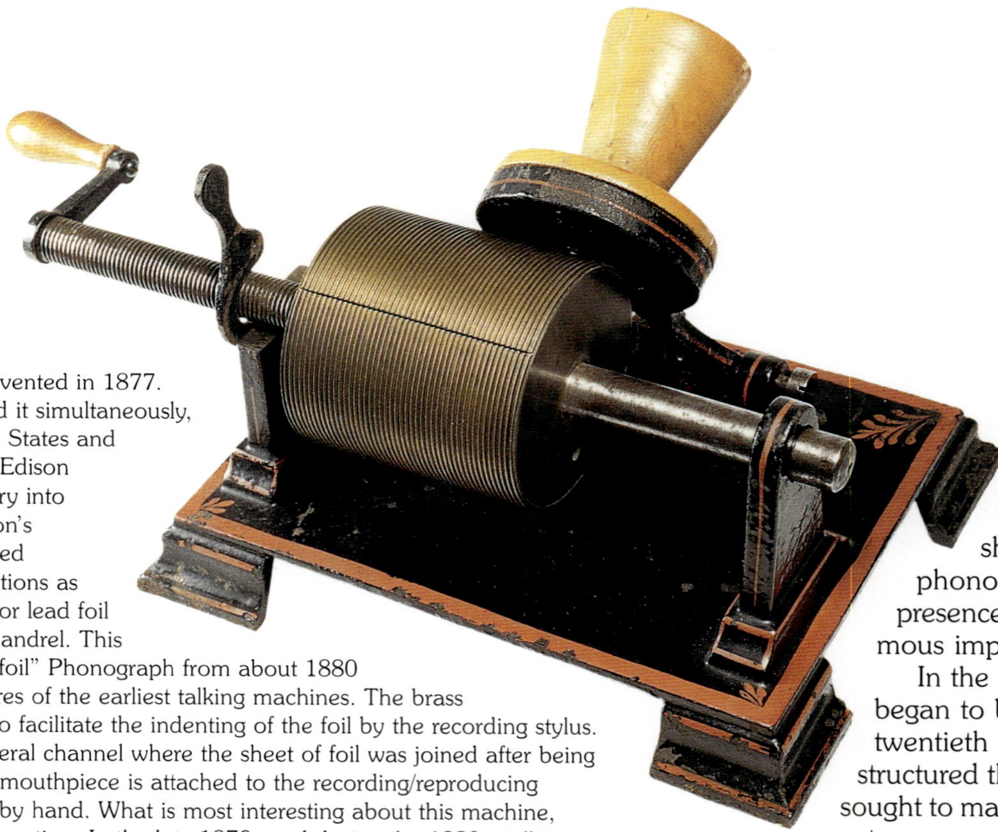

0-1. The phonograph was invented in 1877. Although two men envisioned it simultaneously, Thomas Edison in the United States and Charles Cros in France, only Edison succeeded in putting his theory into practice. The result was Edison's "Tinfoil" Phonograph, so called because it stored sound vibrations as indentations in a sheet of tin or lead foil stretched over a cylindrical mandrel. This European version of the "Tinfoil" Phonograph from about 1880 displays all the primary features of the earliest talking machines. The brass mandrel has a spiral groove to facilitate the indenting of the foil by the recording stylus. The mandrel also shows a lateral channel where the sheet of foil was joined after being applied. The turned wooden mouthpiece is attached to the recording/reproducing mechanism. Motive power is by hand. What is most interesting about this machine, however, is the bright red decoration. In the late 1870s and during the 1880s, talking machines were viewed as scientific instruments. Through most of the 1890s, talking machines were considered to be office or household appliances. For about twenty years after its inception, little effort was expended to make the phonograph look attractive (apart from some gold pinstriping of metal parts). The machine shown here is a rare exception, using bold color in broad strokes to embellish the black castings. (Value code: A)

It is such an ordinary thing to be surrounded by recorded music in our everyday lives, we may never imagine how extraordinary it once was to have a phonograph in the house. Although familiarity has bred a casual relationship between twenty-first century man and the phonograph, CD player, and jukebox, the physical presence of recorded music machines once carried enormous impact.

In the decades following 1906, phonograph cabinets began to be actively designed. During the course of the twentieth century, these designs followed three loosely structured themes. The first theme included designs which sought to make the phonograph inconspicuously blend with contemporary furnishings. Another theme was embodied in attempts to *disguise* the phonograph as some unrelated object (such as a table or sewing machine). Making up the final theme were designs whose sole purpose was to shatter conventions. Often, combinations of these themes can be seen in the same phonograph. There are many "multi-theme" phonographs in this book.

0-2. From 1888 until 1895, there was only one practical talking machine available in significant numbers. Despite its heavy motor and cumbersome batteries, the venerable Edison Class "M" Phonograph was viewed in its time as a scientific marvel; an instrument that performed the seemingly impossible. Little thought was given toward embellishing its appearance. To be sure, the Class "M" was available with "Extra Fancy Gilding" (more elaborate pinstriping) as well as nickel-plated and gold-plated playing mechanisms. Such specially decorated pieces survive in small numbers, but especially noteworthy is this example (No.9438). The upper casting was festooned with additional stripes, floral designs, and curlicues. This machine was a rare early example of infusing a cumbersome, utilitarian apparatus with decorative flair. *Courtesy of Nina and Dave Heitz.* (Value code: VR)

During the approximately ten year period prior to 1906, phonographs, or talking machines, were good examples of form following function. The machinery was sold in plain oak boxes, along with horns finished primarily in black. The purpose of these instruments was to be "heard and not seen." They were considered "scientific" devices transposed to the home. Occasionally, the phonograph would be placed in a purpose-made cabinet designed to harmonize with the furniture, but these wooden enclosures were mainly intended to provide record storage. Visual embellishment of the machines was customarily limited to multi-colored decals and a few pinstripes. If phonographs at the turn of the twentieth century had a "style," it was that of attempted conformity.

0-3. The Douglas Phonograph Company of 89 Chambers Street, New York City, was a prominent retailer of talking machines. It advertised regularly and became well known for its unusual phonograph and record cabinets. This "Perfection Pivot Cylinder Record Cabinet" in "Marquetrie" was illustrated in the August 1905 issue of the *Talking Machine World*. It was offered in five finishes: "Golden Quartered Oak, Mahogany, Rookwood, Vernis Martin, [and] Marquetrie." Within this cabinet is a nickel-plated Edison "Home" Phonograph. Measures 26 1/2" wide x 15" deep x 50 1/4" high including lid. *Courtesy of Nina and Dave Heitz.* (Value code: VR)

First, let's examine the physical design of *cylinder-playing* models. The wooden cabinets of the earliest cylinder phonographs featured veneered bent-oak covers which mimicked the appearance of a common household item: the sewing machine. The acceptance of the latter invention into American homes had been overwhelming. The early cylinder phonograph was not above playing the impostor to bring musical "art" into the home. Phonograph manufacturers recognized that certain fashions in home decoration must be observed if their products were to be embraced by the public. The light oak finishes of the 1890s gave way to a "green oak" finish around 1900. By 1905, a much darker oak finish had become the industry standard, along with special-order mahogany cabinets.

The earliest *disc* phonographs were different from cylinder models in that they resorted to no disguises. Invariably, disc instruments prior to 1906 consisted of a roughly square cabinet (displaying differing degrees of ornamentation) with a horn appended. This defiant flouting of domestic decorum was given an unexpected boost by an oil painting of a little dog, named "Nipper," listening to "His Master's Voice." By 1902, this image had become familiar to millions, along with the concept of a phonograph horn attached to a wooden box in one's parlor.

0-5. Close-up showing the lush detail and rich color of the "Marquetrie" finish. *Courtesy of Nina and Dave Heitz.*

0-4. Douglas described this "Perfection" cabinet as "The most practical Cabinet constructed for this purpose. Drawer shelves revolve. No binding due to warping. All records fully exposed to view when drawer is open. Hollow turned wooden pegs firmly screwed to drawer." Each of the five drawers contains 28 wooden pegs (140 total). *Courtesy of Nina and Dave Heitz.*

By late 1904, the horn had become the focus of both cylinder and disc phonograph design. Cabinets could hold just so much gingerbread, but horns could be enlarged, creatively shaped, painted or plated, and could serve as a "canvas" for artwork applied to their surfaces. Indeed, by 1906 the horn had assumed such focus that a thriving aftermarket developed for horns of every color, size, shape and material imaginable.

So prominent became the once-utilitarian horn that, a century later, it still symbolizes the heyday of the early phonograph to millions of people across the globe. The glorification of the horn may be viewed as the first aesthetic "styling" of the phonograph. It was bold marketing; an "in-your-face" expression that the talking machine was here to stay. The fact that many of these colorful horns were aftermarket items, selected and purchased directly by phonograph customers, made the next development even more dramatic.

Left:
0-6. This Douglas cabinet appeared in the July 1906 issue of the *Talking Machine World.* The example shown here consisted of a Victor "VI" mechanism in a Louis XV style cabinet that predated by four years Victor's similarly decorated Vernis-Martin Victrolas. *Courtesy of the Domenic DiBernardo collection.* (Value code: VR)

Right:
0-7. When the articulated doors of the gilded Douglas cabinet were opened, sliding record storage racks could be accessed. *Courtesy of the Domenic DiBernardo collection.*

0-8. A detail of the Douglas shows the professional handiwork that embellished what was essentially a simple cabinet. This is one of the earliest examples of period design applied to a phonograph cabinet. *Courtesy of the Domenic DiBernardo collection.*

When the Victor Talking Machine Company introduced the *internal horn* "Victrola" in August 1906, the talking machine industry was set on its ear. It seemed that the "Victrola" concept was exactly what the public now wanted: a talking machine that looked like a piece of furniture, with no visible horn.

The genius of Victor's marketing was that the company sought to define exactly what comprised "a piece of furniture." While other companies, particularly Columbia, began the wholesale adoption of *existing* traditional furniture forms for talking machine cabinets (notably upright and grand pianos, desks, and tables), Victor took a different course. The company's advertising successfully created public acceptance of the distinctive Victrola cabinet as a piece of furniture *unto itself*. This disinclination on Victor's part to subordinate the Victrola's identity achieved a hegemony that lasted 15 years.

0-9. The Victor Talking Machine Company had a British cousin, known first as the Gramophone Company, Limited, and much later as EMI. In 1905, this British firm introduced a truly extraordinary disc talking machine known as the "Melba" (shown here with matching stand), in honor of the renowned soprano Nellie Melba. The iconoclastic influence of Art Nouveau is apparent in the detached pillars and the molding of the record stand on which the machine is sitting. However, the overall look is traditional, more grounded in the architecture of heroic monuments than revolutionary trends in furniture design. *Courtesy of the Sanfilippo collection.* (Value code: VR)

Indeed, numerous other talking machine manufacturers adopted as many of the Victrola's features as possible in order to gain public acceptance of their own products. Such was the success of Victor's advertising that the Victrola's familiar lid, doors, and "winged" corners became American icons. Columbia's first internal horn model, the "Symphony Grand" of 1907, might have been a more stimulating visual concept (it resembled an upright piano), but it was far less imitated than Victor's Victrola.

After the appearance of the Victrola, everything changed in the phonograph industry. Each company took steps to standardize its playing mechanisms while offering a widening range of cabinet styles. The quality of sound reproduction took a back seat to the visual appearance of the cabinet. In fact, sound quality actually suffered in comparison to the external horn models.

Companies vied with each other to persuade the public that their cabinets were the most "artistic" in the industry. Edison even employed a noted interior designer, Elsie de Wolfe, to endorse the style and construction of phonograph cabinets bearing his name. The excesses of phonographic cabinet design easily surpassed the horn embellishments of the preceding period. Likewise, a secondary industry sprouted up, focused on phonograph cabinet manufacture. This activity was centered in Grand Rapids, Wisconsin, and made the area a boomtown for several years.

As time passed, phonograph cabinet configuration shifted from the original "upright" look first popularized by the Victrola, to a lower, wider, "console" style. Although both upright and console models were offered in the early 1920s, they were made obsolete in 1925, when technological advances appeared which would influence phonograph cabinet design for the next several years.

1925 was the year when electrical recording and mathematically correct "exponential" internal horn design were introduced. For the first time in years, the public was astounded by what it heard. Not since the days of the traveling phonograph showmen of the 1890s, when crowds had gasped in awe at the noises emitted by cylinder talking machines, had people been treated to such a startling development in sound reproduction. Fidelity was so much improved that the older cabinet machines with haphazardly designed internal horns were forever relegated to obsolescence.

A glance at a phonograph cabinet could tell anyone whether the phonograph in question employed a "modern" or "old fashioned" reproducing mechanism. The long, precisely engineered exponential horns were *folded* into cabinets that were necessarily capacious. Immediately, the predominant cabinet style of the mid-to-late 1920s became large and boxy with a flat lid and a dark lacquered finish: the signature styling of Victor's extremely popular "Orthophonic" Victrolas. Even lesser instruments lacking exponential horns sometimes masqueraded in "Orthophonic-like" disguise. Once more, Victor cabinet styling led the industry, but unlike the example of the first internal horn instruments, this time sound quality did not suffer as a consequence.

The already roomy cabinet of the "Orthophonic" Victrola easily could be made large enough to incorporate a radio receiver. Radio-phonograph combinations became popular in the late 1920s, a marriage which would persist in modified form to the present day. The phonograph would never again enjoy the independent identity it had prior to the advent of radio-phonograph combinations, but trends in interior design would continue to shape its appearance.

The Queen Anne styling of the 'teens gave way to the Louis Quatorze or Italian Renaissance look of the late 1920s. This yielded to Art Deco in the 1930s, with its shiny surfaces and parallel lines, integrated with majestic curves. This vivid style gave way to the more subdued Industrial Modern fashion of the 1940s. In that trend, browns and earthy hues dominated within a rectangular format, as if to reinforce the technical nature of objects. This use of unadorned, monochromatic surfaces stressing simple lines evolved, thereafter, into the sharp angles and pastel colors associated with Atomic Modern in the 1950s. The look remained uncomplicated, but overtly splayed legs and the asymmetrical placement of features rattled at the cage of Eisenhower conformity. Long, low "consoles" appeared during this period, often featuring the "blond" wood finishes of "Contemporary" furniture.

Pop Modern of the late 1960s introduced bright, often garish color schemes, and skewed geometrical "op art" decoration. Reflecting the revolutionary foment of the period, record players became more obtrusive, largely constructed of plastic and less tied to trends in wooden furniture. Widespread use of plastic continued into Disco Modern of the mid-1970s, but iconoclastic visual statements (such as dizzying op art decoration) were replaced by an air of cheap luxury. Disco Modern stressed orderly geometric forms in phonograph and speaker design: egg shapes, cubes, domes, and spheres. Plexiglas turntable covers, and black, white or chrome metal finishes framed in fake wood represented the phonograph's final glimmer of style.

Each of these periods in America exploded with individualistic expressions of popular fashion, though undeniably driven by a commonly shared vision. Hence, we can recognize the dashboard dials and controls of a 1950s automobile, fresh out of Detroit, mirrored in the knobs and levers of a hi-fi set from the same period. Moreover, this "dashboard aesthetic" was to be found reflected everywhere in American society: in dishwashers, stoves, and television sets.

0-10. The cabinet of this 1906 "Harvard" disc talking machine (Style "No.4," $17.40) with its crazily distorted Ionic columns, looked like a refugee from *The Cabinet of Dr. Caligari.* A definite Teutonic flavor invested at least two of the machines Sears, Roebuck and Company sold under the "Harvard" brand (note the "form follows function" carrying handle), but this model seemed to exhibit the influence of German Expressionism. Such a departure from the parallelism of external horn talking machine cabinets was extremely rare.

The physical appearance of the phonograph reacted to styles, trends, and tastes throughout the first eight decades of the twentieth century. The phonograph was shaped by the world around it, and by the shared understanding of the people who inhabited that world. Conventional wisdom holds that the value of antique phonographs lies in their ability to reflect the auditory history of society. In this book, we hope to present the phonograph's visual contribution (and *evolution*).

As this book is being written, at the dawn of the twenty-first century, sound reproduction equipment has come full circle. Today's audio systems are perfect examples of form following function. The squat metal and plastic boxes finished in matte black, along with ever-smaller speakers likewise unobtrusive and finished in black, are meant to be "heard and not seen."

Typically, today's CD players, receivers, DVD players, VCRs and cassette players are hidden away in purpose-made cabinets sold as "entertainment centers." The utilitarian electronic components that are housed in these furniture-like shells are designed solely for optimum sound reproduction. They are blatantly presented as technical equipment. Visual embellishments are limited to multi-colored LED's and the imaginative arrangement of controls. Though the record player is effectively obsolete, replaced in the past 15 years by the CD player, looking at today's sound equipment seems like old times. The phonograph began its long, strange trip across the twentieth century as a plain wooden box.

CHAPTER ONE, 1907–1919: THE PHONOGRAPH IN DISGUISE

1-1. The "Symphony Grand" of 1907 was the Columbia Phonograph Company's first disc talking machine with a concealed horn. Barely concealed, you might say, since behind the decorative molded composition grille was a conventionally-belled metal horn, grain-painted mahogany to match the wood of the cabinet. In fact, all the internal workings of this machine were virtually identical to equipment then being sold on such deluxe exterior horn Disc Graphophones as the "BD" and "BY." It was the cabinet of the "Symphony Grand," in imitation of an upright piano, which trumped Victor's first internal horn machine, the unassuming "Victrola" of 1906 (although the Victrola was ultimately more successful). It may strike us as odd that the public would have accepted such a peculiarly truncated (31 3/4" wide x 23 1/2" deep x 52 1/4" high) and expensive ($200.00 in 1907) effigy of a piano. However, it helps to remember that "upright" pianos were then known in a number of types and sizes. Most common, of course, were the weighty behemoths found in the home. Yet, there were smaller and significantly narrower coin-operated pneumatic pianos (some with as few as half the keys of a full-sized "88") to be seen in public places. Within the context of its day, the "Symphony Grand" would not have seemed unfamiliar. There were two distinct models of the "Symphony Grand": one with side-opening cupboards to store discs, and one (shown) with sliding record racks which withdrew from the sides of the cabinet. U.S. design patent No.38,669 awarded to John Herzog. *Courtesy of the Domenic DiBernardo collection.* (Value code: D)

1-2. The fall board of this instrument hid the turntable, tone arm and soundbox.

Right:
1-3. Victor's Victrola was the innovation that spurred the history of phonograph cabinet design. The initial model of 1906 rapidly evolved into the style of instrument shown here, the "VTLA" (short for "Victrola") of 1907. In 1908, the "VTLA" became available in Circassian walnut, an exclusive and expensive option (adding $50.00 to the $200.00 price of the machine). This example perfectly displays the fluid grain patterns that made Circassian walnut coveted by many, though affordable to few. *Courtesy of John W. Hess.* (Value code: D)

Far right:
1-4. By 1910, the styling of the Victor Victrola had already become immediately recognizable to the buying public. Columbia Grafonolas, however, had yet to assume a visual persona. This "Mignon" Grafonola may have been a deliberate attempt by the Columbia to offer an antithesis to the curvaceous lines of the Victrola. Appearing to be a simple chest mounted on four squat legs, the severe angularity was somewhat softened by the recessed door panels and frieze work on the front. The cabinet was available in mahogany (shown), or oak in golden or mission finish. Original price was $150.00, which was $50.00 less than Victor's popular Victrola "XVI." However, the boxy appearance of the "Mignon" did not endear itself to the public, and the model was a poor seller. *Courtesy of William G. Meyer.* (Value code: G)

1-5. The "Mignon" Grafonola hid needle cups in the left-hand rear corner beneath an elegant hinged cover. Unfortunately, this model suffered from a deeply-recessed playing compartment similar to that which had plagued the very first version of the Victrola. By 1907, Victor had realized that this flawed design impeded access to the turntable, and had made the playing compartment shallower, and introduced its signature domed lid. American Graphophone unfortunately failed to profit from Victor's experience, and thus was doomed to learn the same lesson first hand. *Courtesy of William G. Meyer.*

Right:
1-6. In 1910, entrepreneur Morris Keen had big changes in mind for his Keen-O-Phone Company of Philadelphia. His family had been involved with selling phonographs and records for years, and Morris was preparing to transform his musical instrument business into eponymous talking machine manufacture. His Keen-O-Phone disc instruments would prove true to the slogan under which he first marketed them: "New Idea." The machines were truly original in both cabinet design and mechanics. Flagship of the line was the "Keenolophone," (U.S. Design Patent No.41,402, filed March 7,1911) also called the "XXX" or "30," which was introduced in 1911 (see illustration 1-8). However, before the "Keenolophone" was put into production, this earlier version of the machine (probably 1910) was constructed. In appearance, it suggested the same general outline of the later model, but differed significantly in cabinet construction and detail. On March 17, 1910 Morris Keen had applied for a design patent (granted as No.40,699) which exhibited substantial elements of the machine shown here. *Courtesy of Nina and Dave Heitz.* (Value code: VR)

1-7. Opened, this very early "Keenolophone" displays record niches in the lower half, undoubtedly designed for use with album books. Above, a hatch door opened to allow access to the playing controls. Since the horn of this instrument comprised the lid of the cabinet, it was necessary to keep the lid closed in order to complete the sound passage. The cabinet has a distinctly "Empire" look that would not be seen in Keen-O-Phone's catalogued models. *Courtesy of Nina and Dave Heitz.*

1-8. Although stencil-decorated metal brackets were common on acoustic talking machines prior to 1920, delicate hand-painting was rare. The tiny pink blossoms sported by this very early example of a "Keenolophone" (1911) were rapidly abandoned as the Keen-O-Phone Company adapted its instruments to mass production. *Courtesy of Brady Jeffcoat.*

1-9. Looking more like the nose of a post-World War II automobile than a 1911 cylinder phonograph, the U.S. "Peerless" was distinguished by a truly original grille panel, within the context of conventional "Victrola-inspired" cabinetry. The visual flux of the front panel suggested the movement of the sound as it swelled into the room. Harry McNulty was granted a design patent (No.45,467) for the "Peerless" cabinet, though he did not apply for it until February 7, 1914 – after the U.S. Phonograph Company had already ceased operation. *Courtesy of Walt and Ellen Bellm.* (Value code: VR)

1-10. The U.S. Phonograph Company of Cleveland, Ohio, made cylinder talking machines between 1910 and 1913. This period represented the final chapter of competition in the cylinder machine market. Thereafter, only Edison would continue to promote the cylinder record format. Through the efforts of the U.S. Company, which produced innovative, eye-catching equipment, the cylinder phonograph did not go gentle into that good night. *Courtesy of Walt and Ellen Bellm.*

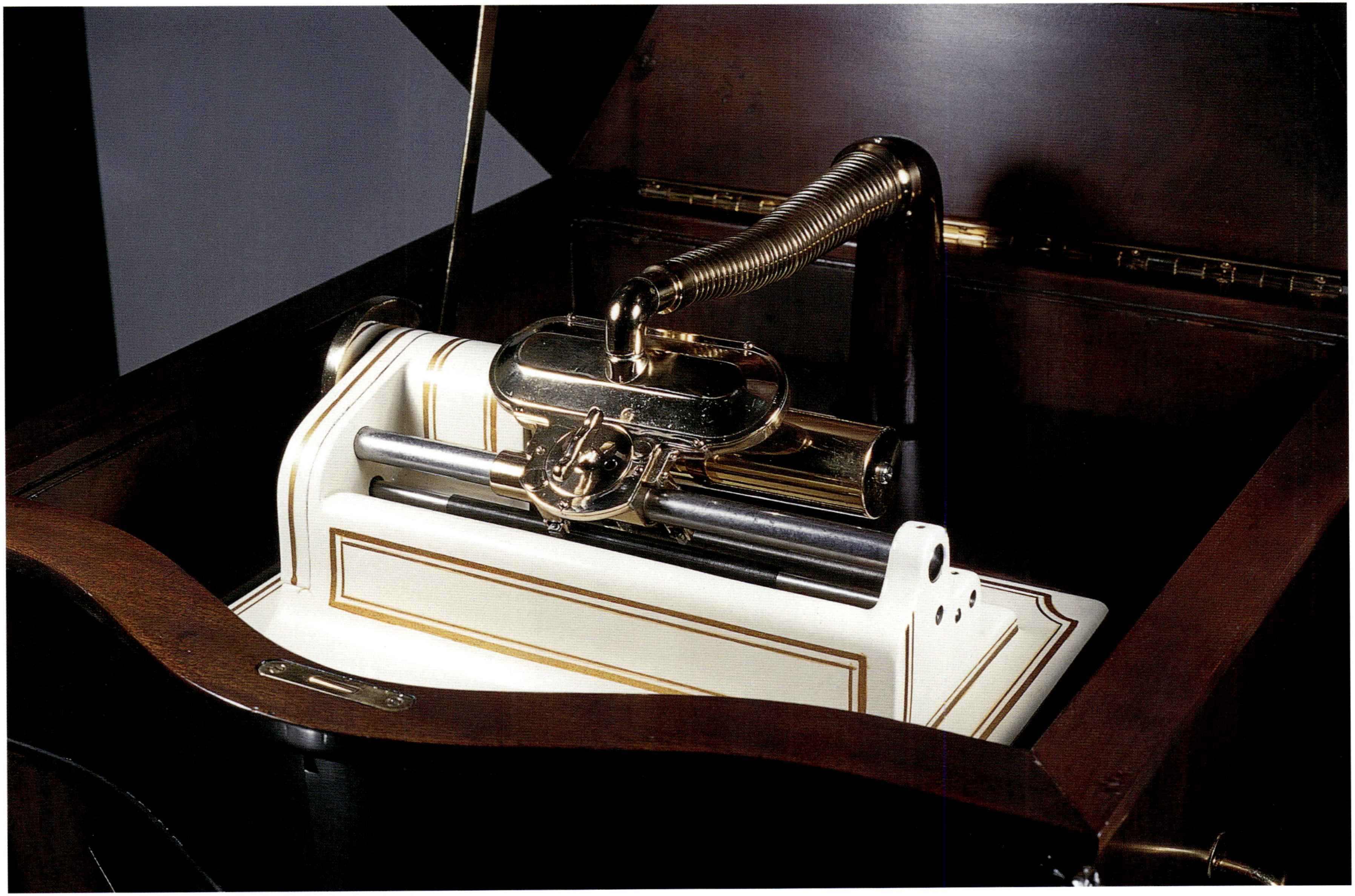

1-11. The stunning interior of the "Peerless" featured gold-plated hardware on a white enameled casting. Traditional black-painted phonograph castings may have been more practical, considering the grease and oil that was likely to accumulate on them, but the luminous playing mechanism of the "Peerless" uniquely levitates within the dark interior of the cabinet. *Courtesy of Walt and Ellen Bellm.*

1-12. *Courtesy of Walt and Ellen Bellm.*

1-13. Within the first five years of the twentieth century, it was recognized that the phonograph horn, regardless of how it was embellished, was not aesthetically pleasing to people of refinement and culture. The Hymnophone of 1904 hid *most* of the horn within a small cabinet, though the bell remained quite conspicuous. Not until the Victrola of 1906 was the horn as well as the playing mechanism completely hidden. This German phonograph of the 1910-1912 era appeared to be a mahogany box which might have contained a smoking set or liquor decanters. Yet, the mechanism within took horn camouflage in a new direction. *Courtesy of Garry James.* (Value code: VR)

1-14. Opening two sides of this "hornless" phonograph reveals a playing mechanism with a typically Germanic appearance. Note, at first glance, no horn is visible. *Courtesy of Garry James.*

1-15. A closer look reveals a horn made of glass. This indeed was diminishing the horn in extreme, yet simple, fashion. However amusing this design might seem a century later, this phonograph accomplished its purpose efficiently in a limited space (approximately 18" x 18") and utilized the most appropriate materials. *Courtesy of Garry James.*

1-16. In a class of their own were the Victrolas supplied in the special-order Vernis-Martin finish, first offered in August 1910. Such embellishment doubled the price of a Victrola "XVI" from $200.00 to $400.00. Gold leaf, additional moldings, and hand-painted scenes in the Louis XV style enhanced this example, No.56789, built in early 1911. *Courtesy of the Sanfilippo collection.* (Value code: VR)

1-17. A close-up of one side of the Vernis-Martin Victrola reveals the delicate shading and overall care that was invested in this extraordinary finish. Vernis-Martin finishes were especially susceptible to deterioration from rapid changes in temperature, reducing over the years an already select few to a mere handful of survivors. *Courtesy of the Sanfilippo collection.*

Right:
1-18. Even the lid of a Vernis-Martin instrument was embellished with pastoral scenes and floral decoration. Victor legend tells of a sixth-floor corner of a factory building where old-world craftsmen labored for weeks on each Vernis-Martin cabinet. Until discontinued in October 1917, this style represented the apogee of Victor craftsmanship. *Courtesy of the Sanfilippo collection.*

1-19. In 1912, Columbia introduced one of its most ambitious Grafonolas, the "Grand." Although the nomenclature revived a term which had been used some ten years earlier to describe Columbia's deluxe cylinder talking machines ("Grand" Graphophones), the $500.00 price of this Grafonola represented a new height in regularly catalogued instruments. In this view, the Columbia "Grand" Grafonola could be mistaken for a baby grand piano, except that the dimensions of the body measure only 32 1/2"wide x 41 1/2" long. U.S. Design Patent No.44,553 awarded to Clinton E. Woods. *Courtesy of the Domenic DiBernardo collection.* (Value code: D)

1-20. Unlike the earlier "Symphony Grand," this instrument did not suggest the presence of a keyboard. Instead, sound-regulating louvers occupied the front of the cabinet, along with two cupboards for record storage. *Courtesy of the Domenic DiBernardo collection.*

Below:
1-21. In the plush interior of the Columbia "Grand" Grafonola, gold-plated hardware wallowed in rich green velour. The hooded electric light in the foreground was triggered by uncovering the playing compartment. *Courtesy of Brady Jeffcoat.*

1-22. Columbia published a posh sales book to promote the "Grand."

1-23. The "Grand" was illustrated in tony surroundings, while its advantages were extolled. The machine featured electric motor drive and electric illumination. All metal parts were finished in gold.

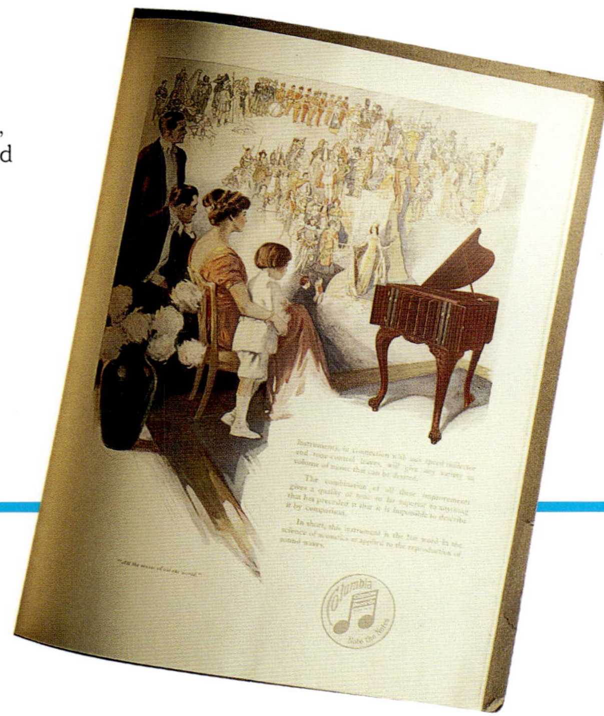

1-24. Shamelessly copying a familiar marketing motif of the Victor Talking Machine Company, the Columbia "Grand" Grafonola was illustrated surrounded by the many performers whom it brought into the home.

1-25. A little known accessory of the Columbia "Grand" was a "piano bench" styled to match the instrument. *Courtesy of the Domenic DiBernardo collection.*

1-26. The bench was in fact a record storage cabinet, with wooden files that held the discs between indexed paper separators. *Courtesy of the Domenic DiBernardo collection.*

1-27. When the Edison Diamond Disc Phonograph line was being developed in 1912, the company recognized that a model was needed at the upper end to epitomize the quality and style that the line was meant to evoke. Joseph Rubino, a prolific cabinet designer for Edison, created an ornate prototype which blended classical elements with a distinctive "old world" execution, known as the "Louis XVI" (U.S. Design Patent No.43,161). A mahogany version was offered for $400.00, but the *pièce de résistance* was this model in Circassian walnut priced at $450.00. The Edison "A-450" (later "B-450") was produced from late 1912 until mid-1916, when World War I curtailed the supply of Circassian walnut. Built in small numbers, the "A/B-450" was last offered in the September 1918 catalogue, aptly coinciding with the end of the "old world" that inspired this beautiful cabinet. *Courtesy of the Sanfilippo collection.* (Value code: VR)

1-28. The Edison Diamond Disc Phonograph "A-290" represented the highest level of decoration available on an otherwise uninspiring "Sheraton" cabinet. The "A-290" ($290.00) featured marquetry on three sides and the lid. Introduced in 1912, this cabinet was protected under U.S. Design Patent No. 43,160 awarded to Joseph Rubino. *Courtesy of Nina and Dave Heitz.* (Value code: VR)

1-29. The lid of the Edison "A-290" was distinguished by patterned veneer and marquetry. Public acceptance of this expensive model was extremely slow: only 167 had been sold shortly before it was discontinued in February 1915. *Courtesy of Nina and Dave Heitz.*

1-30. The Edison Company offered its Diamond Disc Phonographs in a variety of expensive "period" cabinets. The Model "A-425" in Circassian walnut shown here was one of the most posh. It was introduced in 1912 at $425.00, lasting only about two years in the catalogue because of poor sales. In this luxurious instrument, the Louis XV style has been little compromised, and none of the extraneous design elements that Edison and other firms employed to "pretty up" their phonograph cabinets have been tacked on. *Courtesy of Walt and Ellen Bellm.* (Value code: VR)

1-31. The veneer of the "A-425" was artistically matched to create the impression of a shimmering pool of soft colors, into which the eye is drawn and the imagination submerges. *Courtesy of Walt and Ellen Bellm.*

1-32. The classically-inspired terpsichore of Isadora Duncan is reflected in this 1912 Victrola "IX" with unique decoration. The border at the top of the side panel may be unexpectedly Early American, but the female figures, clad in girdled *chitons*, are lithe interpretations of ancient Greek vases. The meaning of the names, in pseudo-Greek characters, is unclear. *Courtesy of Fred Guenther.* (Value code: VR)

1-33. On the rear panel of the "IX," the dancers hold wreaths. A wreath also crowns the flat of the lid, though the geometric painting of the sloping sides repeats the influence of American folk art. The decoration appears to have been underlain with gesso, and the surfaces are textured or in low relief. The composition and execution of the figures seems inspired by Burne-Jones, but the exact purpose behind this carefully decorated Victrola is a mystery. *Courtesy of Fred Guenther.*

1-34. Though Edison floor-standing phonographs from the 'teens and 1920s were customarily identified with a certain school of furniture design (such as Chippendale, Sheraton or Hepplewhite), their cabinets were most often patchy, inaccurate adaptations of these schools. Edison's purest representation of a recognized style was the "Arts and Crafts" Amberola "IV" cylinder Phonograph of 1913. It was designed with the principles of Gustav Stickley or Elbert Hubbard in mind. Both men were influential in the American "Craftsman" movement of the early twentieth century. *Courtesy of the Domenic DiBernardo collection.* (Value code: C)

Right:
1-35. Ironically, the Amberola "IV" was a complete failure in the marketplace. Its faithfulness to the stout, earnest, mortise and tenon look of the American Arts and Crafts (or Mission) style proved anathema to the public. This was a rather unusual reaction, considering that Mission furniture was well known at the time. After a couple hundred had been produced, the machine was quickly withdrawn. *Courtesy of the Domenic DiBernardo collection.*

1-36. Of course, it didn't help that the workings of the Amberola "IV" were a hodgepodge, consisting of obsolescent mechanics that Edison was trying to exhaust. Yet, if the cabinet had been "sweetened up" with a few silly curlicues or some inappropriate appliqués, as Edison and other phonograph companies frequently did, the "IV" might have been made more palatable to public taste. *Courtesy of the Domenic DiBernardo collection.*

1-37. The Victrola "XVIII" occupies a special niche in phonographic history. The top of the catalogued Victrola line at $300.00, it was offered for only 18 months (June 1915-December 1916). The "XVIII" featured bombé surfaces, elaborate moldings, and a pleasing compactness (U.S. Design Patent No.47,379 awarded to Eugene T. Kieffer). Normally, this model was offered in "V-grained" mahogany, but this example was given a very early (for the phonograph industry) version of an oriental motif. In this case, the Victrola was painted matte black with gold highlights, and featured Japanese designs hand-painted on the cabinet panels and lid surfaces. (Value code: VR)

Right:
1-38. The interior of the Japanese Victrola "XVIII" shows the electric light supplied with electric-motored versions. Note the "Electrola" nomenclature on the lid decal.

1-39. A side view of this extraordinary Victrola "XVIII" shows the panel painted in the style of a Japanese screen and "aged" around the edges. The original price of this particular example is unknown, but for a special order finish with electric motor ("Electrola") option, this instrument was surely flirting with $500.00, if not more.

1-40. A close-up of the lid shows the geometric and delicate floral decorations that distinguish this cabinet. With the gradual demise of the earlier Vernis-Martin finishes by the end of 1917, one wonders if the same artisans were responsible for the striking Japanese designs found on this instrument.

1-41. In the 'teens, Columbia pioneered a successful line of (internal horn) Grafonolas masquerading as ordinary household tables. The smallest of these was the "Princess," a Queen Anne style end table (here shown in mahogany). U.S. Design Patent No.43,294 awarded to S.M. Wirts. (Value code: H)

Right:
1-42. The Grafonola tables were designed for dual functionality. The narrow door on the right side allowed access to the playing controls while the tabletop remained undisturbed. Alternately, the top could be angled back for greater convenience. The louvers at the front of the machine regulated volume.

1-43. The Grafonola "Baby Regent" was a Queen Anne table of both style and substance. The flow of the lines and solidity of the construction were exceeded only by the cleverness of function. The talking machine mechanism was contained within a smoothly gliding drawer. The mouth of the internal horn was opposite the face of this drawer. The example shown here was veneered in elegantly figured quartered oak. A version without a lower shelf is also known. U.S. Design Patent No.43,293 awarded to S.M. Wirts. *Courtesy of Freddie and Jeanette Blair.* (Value code: F)

1-44. The Grafonola "Regent" was the most imposing member of Columbia's line of talking machines disguised as furniture. The cabinet, though it suggested a desk, was poorly proportioned for any use other than a record player or a surface on which to arrange household bric-a-brac. *Courtesy of the Domenic DiBernardo collection.* (Value code: F)

1-45. The appropriately-named "Tabe-a-nola" was produced in the late 'teens by an independent firm, competing in the Grafonola "furniture format." Even Columbia's "phonograph-in-a-drawer" concept has been copied, though the overall design has been altered from Columbia's signature Queen Anne to a more rectangular Hepplewhite look. *Courtesy of Freddie and Jeanette Blair.* (Value code: VR)

1-46. Aeolian-Vocalian's elegant "900-2," suggesting the Hepplewhite style, was the type of instrument the firm began offering when it entered the table-phonograph war in the mid-'teens. Here, the entire tabletop lifted to reveal the playing mechanism (thus limiting the instrument's value as a table). The mouth of the internal horn was uncovered by opening a door in the rightmost face panel. Note that the "900-2" was designed to sit flat against a wall. *Courtesy of Freddie and Jeanette Blair.* (Value code: F)

1-47. The Aeolian playing compartment occupied the left third of the interior, while a felt-covered surface to the right was intended for temporarily arranging records. *Courtesy of Freddie and Jeanette Blair.*

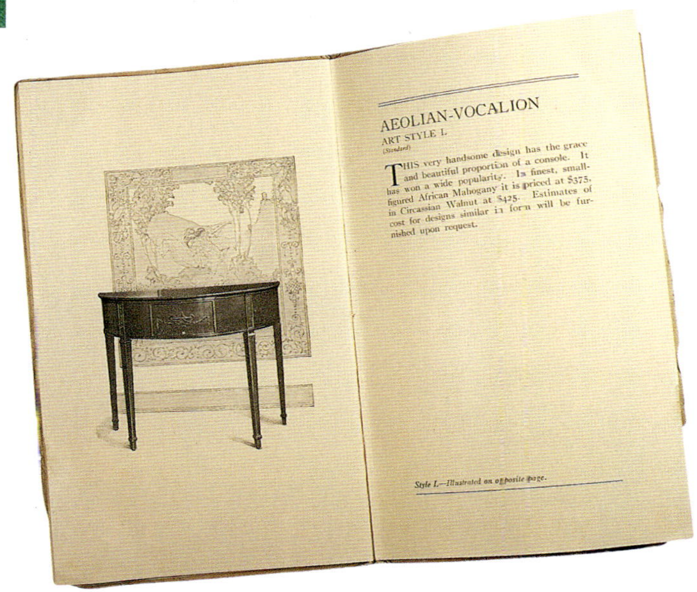

1-48. An Aeolian catalogue, circa 1916, illustrated the "Art Style L," an earlier version of the "900-2." Note the lofty $375.00 price. *Courtesy of Freddie and Jeanette Blair.*

1-49. In 1917 Victor inaugurated a line of pricey "period" Victrolas, of which this is an example of the William and Mary style (upright). The turn-of-the-XVIII-century elements usually incorporated into William and Mary designs are seen here, most notably the flat (as opposed to recessed) surfaces, drop pulls and under-bracing of the legs. Historically correct, but less common in reproduction, are the spiral corner pillars and legs, which lend a surprisingly Byzantine flavor. *Courtesy of Nina and Dave Heitz.* (Value code: VR)

Opposite page:
1-50. The upright "period" Victrolas, ordinarily priced between $500.00 and $700.00, sold in very low numbers. An unusual feature is the use of double needle receptacles. Note the small drawers which caught used needles. Beyond this, the interior is virtually identical to other Victrolas of the time. *Courtesy of Nina and Dave Heitz.*

SOUTH NINTH ST.
TALKING MACHINE SHOP
JOSEPH CAROLEI
VICTROLAS
PIANOS
Victor
VICTOR TALKING MACHINE CO.
V-WILLIAM & MARY No.O.
516

Far left:
1-51. In September 1917, Edison introduced its version of the William and Mary style as Diamond Disc Phonograph, Model "W-250." The cabinet design was by Raimondo C. Aimone, U.S. Design Patent No.52,721. The Aimone Manufacturing Company of New York City produced cabinets for a number of different Edison models, but the ultimate expression of the firm's expertise surely must have been this special order "W-250" in Chinoiserie finish. *Courtesy of Nina and Dave Heitz.* (Value code: VR)

Left:
1-52. The whole instrument has been hand-decorated with gilded oriental scenes. The background colors fairly glow. *Courtesy of Nina and Dave Heitz.*

1-53. Intricate decorative patterns were associated with late William and Mary styling. Here, the grille has been brilliantly interpreted in an oriental vein. *Courtesy of Nina and Dave Heitz.*

1-54. Even the top of the cabinet lid has been spectacularly adorned. *Courtesy of Nina and Dave Heitz.*

1-55. In August 1918, during the "period cabinet" craze, the Edison company introduced its XVIII Century English (Adam) "C-450" Diamond Disc Phonograph. The uninspiring rectangular cabinet was relieved only by cartoonish paintings in neo-classical motifs. The aesthetics of the applied decoration aside, it is difficult to imagine the appeal of the price, which rose from $450.00, upon introduction, to $500.00 in December 1919. Compare this overpriced shoebox to the graceful (and less expensive) Louis XV Diamond Disc model shown in illustration 1-30. The lack of public interest was no doubt reflected in a price reduction to $375.00 after March 1922. Yet, the "C-450" intrigues us today with its flatly rendered Revolutionary War era visions of the ancient world. *Courtesy of Gus Schild.* (Value code: F)

1-56. The Brunswick-Balke-Collender Company was a well-known manufacturer of billiard equipment and high quality wooden bar interiors for many years before it began supplying phonograph cabinets to various companies in the 'teens. By 1916, Brunswick was offering its own complete phonographs to the trade. Its entry into the talking machine market was well timed to coincide with the public's preoccupation with cabinetry. Brunswick excelled in exquisite woodworking, of which this phonograph was a prime example. The machine is unmarked, but was illustrated in an advertisement in the October 1918 *Etude*. It featured the oval-shaped horn opening introduced in September 1918, the "Ultona" tone arm and soundbox, as well as an electrically driven turntable. Yet, the large, ornately carved walnut cabinet was by far the most memorable characteristic of this stunning phonograph. *Courtesy of the Sanfilippo collection.* (Value code: VR)

1-57. The signature bombé styling of the Sonora Phonograph Company was most richly embodied in the "Supreme," the firm's top model. Here, that pneumatic look which made lesser Sonora models appear top-heavy finally achieved elegance through the Mannerist attenuation of the cabinet. Beautifully figured mahogany veneer was employed, as well as delicately carved decorative elements. Of all the pretentious ways it might have described the look of the "Supreme," the company, in its 1920 catalogue, chose to say with uncharacteristic bluntness, "Bulged vertically and horizontally." The design patent for this machine was awarded to John Herzog, No.45,460, granted March 24, 1914. *Courtesy of Walt and Ellen Bellm.* (Value code: VR)

1-58. On its more expensive models, Sonora featured wooden tone arms, finished to match the cabinet. As befit its station, the "Supreme" sported an arm of formidable, though practically obscene, proportions. *Courtesy of Walt and Ellen Bellm.*

1-59. A Sonora record cabinet was available to accompany the "Supreme," mirroring the machine in size and ornament. The sheer space requirements of such a pair, not to mention the cost, could only be accommodated in the homes of the wealthy. *Courtesy of Walt and Ellen Bellm.*

1-60. An indexed record filing system assured the music lover ready access to the library of discs contained in the Sonora storage cabinet. *Courtesy of Walt and Ellen Bellm.*

1-61. That at least one Sonora "Supreme" was given oriental decoration is evident in this matching record cabinet. 23" wide x 17 1/2" deep x 50" high. *Courtesy of Nina and Dave Heitz.*

1-62. Sonora's *ne plus ultra* was Circassian walnut. The patterns in the veneer suggest swirling eddies of sand. (Value code: VR)

1-63. Note the more graceful modeling of this Sonora wooden tone arm than the one shown in illustration 1-58.

1-64. The Premier was one of the many brands of talking machine to spring up during the late 'teens and early 1920s. Three slightly different models were catalogued, all suggesting the rough outline of an upright piano. This was a revival of the design premise pioneered by Columbia's "Symphony Grand" of 1907. The fall board concealed the playing compartment of the phonograph, and the sound escaped through the decorative grille-work above. The cupboard below held records. There were three separate design patents for the trio of Premier models, all granted to G. M. Wilson on October 19, 1915. *Courtesy of Steve and Ellie Saccente.* (Value code: F)

Far left:
1-65. If a phonograph can be a lamp, table, or sideboard, why not an "armoire?" In its closed position, one could surely mistake this piece of mahogany furniture for a wardrobe in the gargantuan Queen Anne style popular in the 'teens – except for the suspiciously "Victrola-like" top of the cabinet. *Courtesy of John W. Hess.* (Value code: VR)

Left:
1-66. The "Triumphola," about which little is known, was modeled after talking machines of the "vertical" orientation, such as the Premier or Hawthorne and Sheble's "Starola." The inside of the cabinet was finished in a type of yellow maple that was often employed in piano interiors, and a number of piano producers were known to have constructed phonographs. The horn grille, in the "leaded glass" design commonly used in talking machines during the 'teens, was once backed by gold-colored fabric. The top of the cabinet, while it betrayed a profile reminiscent of a Victrola lid, had no function other than decorative. Cabinet measures 32" wide x 18" deep x 71" high. *Courtesy of John W. Hess.*

1-67. This plate identified the "Triumphola." By constructing the brand name out of a word ending in "ph," and adding the suffix "-ola," three distinct terms were evoked: "triumph," "phone" and "(Victr)ola." Other firms were less clever (or fortunate) in coining their phonograph names, choosing such clumsy sobriquets as "Tabe-a-nola' (see illustration 1-45). *Courtesy of John W. Hess.*

Far left:
1-68. Several companies disguised phonographs as lamps, but perhaps the earliest was the Electric Phonograph Company, 29 West 34 Street, New York City. Advertisements for its products appeared in the August 1916 issue of the *Talking Machine World*. By the time this Phonolamp Model "C" was built, the company had changed its name to the Electric Phonolamp Corporation. Protected by U.S. Design Patent No.55,217, this machine combined style with utility in creating a thoroughly modern and fashionable hybrid of floor lamp and phonograph. *Courtesy of Steve and Ellie Saccente.* (Value code: E)

Left:
1-69. In the late 'teens and into the 1920s, certain firms used wicker cabinets to enclose disc talking machine mechanisms. Among collectors today, the most well-recognized brand is Heywood-Wakefield, but there were others, too. In this unidentified instrument from about 1920, the wicker medium has been made light and fanciful by scrolls and curlicues galore. Observe, also, the musical staff and notes that decorate the horn opening. The mechanical components of this machine are typical of the die-cast arms and soundboxes, and helical gear motors, which many independent phonograph companies employed in their instruments. *Courtesy of Nina and Dave Heitz.* (Value code: F)

1-70. This beautiful phonograph carried no brand name, and was equipped with generic components. Thus, the name of the company responsible for this striking oriental cabinet is lost to us. Note the three natural mahogany rosettes on the grille, and the combination of red, gold and green decorations. *Courtesy of Ron Haring.* (Value code: G)

1-71. The Sonora bombé cabinets of the 1918-1922 period, constructed by the former Herzog Art Furniture Company, were considered stylish at the time on their own merits. However, the owner of this particular machine evidently sought further embellishment of an oriental nature. All painting was done over the existing finish, and carefully applied around some of the hardware. The paintings were professionally rendered, but lack the delicate artistry found on factory-supplied oriental finishes of the period. *Courtesy of the Charles Hummel collections.* (Value code: G)

1-72. Oriental embellishment was by no means limited to the major phonograph brands. The Zemer-phone, by J. Pomerantz, was so enshrouded by decoration that it bordered on garishness. *Courtesy of Brady Jeffcoat.* (Value code: G)

1-73. "Form must follow function," is an oft-quoted phrase, attributed to a variety of schools of design but most commonly associated with the German Bauhaus school. In the case of the Bell phonograph, the aforementioned precept might need to be broadened. In this highly unusual machine, form was the embodiment not only of function but also of the brand name. Here, the bell-like profile can be seen clearly. The Bell was manufactured circa 1920 by the Durable Phonograph Company, Incorporated (with offices in New York City and a factory in Cleveland, Ohio). (Value code: F)

1-74. Structurally, the Bell was little more than a metal acoustic horn to which a motor, turntable, tone arm and soundbox had been appended. It couldn't be considered an *internal* horn talking machine, and it wasn't really an *external* horn one. It was a horn in the shape of a bell, and there was not much more to be said. Interestingly, the Bell's mainspring was wound by a lever arm, a device employed frequently in music boxes but seldom in phonographs. The diaphragm of the soundbox lay horizontally, but the machine played only lateral cut discs. On March 23, 1915, Paul Hunt had applied for a U.S. Design Patent on the design of the Bell, which was ultimately awarded as No.48,249.

Far left:
1-75. This "craftsman-made" disc talking machine and matching record stand emerged from a flight of some skilled carpenter's fancy. The manner in which the stand was constructed to precisely blend with the edges of a table model phonograph, to give the impression of a single unit, was known in the early twentieth century as a "continuation top." The style is American folk art, with its heavy use of prefabricated appliqués, rather than an imitation of any recognized school. *Courtesy of John W. Hess.* (Value code: VR)

Left:
1-76. The maker has copied a Victrola, right down to the sound modifying doors and fixed louvers, which Victor called "sounding boards." Needless to say, the commercial use of these features would have been in violation of Victor patents – and, considering the taste for blood of Victor's legal department, the company might have prosecuted the carpenter responsible for this unique instrument if it had known about him. *Courtesy of John W. Hess.*

Chapter Two, 1920–1929: Toward a New Visual Identity

2-1. Columbia's success in producing a talking machine likeness of a baby grand piano (the "Grand," 1912) inspired other companies to introduce similar instruments. The Fern-O-Grand, manufactured in Cincinnati, Ohio, around 1920, was one such machine. The keyboard area opened to allow the sound to escape from the internal horn, and the impression of a genuine (though diminutive) piano was reinforced by pedals that activated functions of the instrument. *Courtesy of Freddie and Jeanette Blair.* (Value code: F)

2-2. The playing compartment of the "Fern-O-Grand" was equipped with the kind of generic arm and motor that were used by numerous other independent brands during the same period. This particular example has been fitted with an Ellis soundbox, a sound-improving option sold in music stores. *Courtesy of Freddie and Jeanette Blair.*

2-3. The Brunswick "1500" was a formidable mass of carved oak on stout yet graceful legs. This model was not for the faint of heart: it originally sold for $1,500.00. Measures 62 1/4" wide x 26 1/2" deep x 58 3/4" high. *Courtesy of Nina and Dave Heitz.* (Value code: VR)

2-4. Hidden within the Brunswick "1500" were the conventional motor, turntable, and "Ultona" tone arm and soundbox supplied with lesser Brunswick phonographs. Although usually equipped with an electric motor, this particular "1500" was spring-driven. The horn was behind the center grille, and the leftmost section contained record storage. *Courtesy of Nina and Dave Heitz.*

2-6. Opening the doors and lid of the Orsenigo shows the ample record storage and pleated grille cloth of the cabinet. The middle door pulls down much like a desk's writing surface, and slides beneath the grille.

2-5. Little is known about the Orsenigo. Of the few examples that exist, each boasts a unique cabinet which might better harmonize with a theatre or church than a home. During the 'teens, companies began to house phonograph mechanisms in Late Medieval and Renaissance style "credenzas" (meaning a buffet or sideboard), a trend which grew in popularity through the 1920s. In 1925, Victor would dub a substantial model of Orthophonic Victrola the "Credenza." Though Victor's "Credenza" did not employ the prominent carving which had covered earlier machines such as the Orsenigo, Victor used more overt decoration on its "955" and "1050" models. The Orsenigo's imposing cabinet measures 50" wide x 21" deep x 45" high. (Value code: VR)

2-7. A close-up of the decal offers few clues to the Orsenigo's origin. Advertisements for the Orsenigo Company, Incorporated, 110 West 42 Street, New York City, first appeared in the trade journal *Talking Machine World* in February 1922. The mechanism consisted of a Fletcher tone arm and soundbox marked "Orsenigo," and a "Speedrite" electric motor. Like many other phonographs of this period, the Orsenigo was most distinguished by its cabinet, which was of noteworthy quality and design.

2-8. Late Medieval and Renaissance styling conceits were exploited by a number of talking machine firms in the 'teens and early 1920s. These ponderous, deeply carved leviathans were not to the taste, nor within the reach, of most Americans. In this 1923 advertisement, Victor advertised the availability of its "Art Shop" for creating custom "period" Victrolas. One can only imagine the cost of such instruments, when catalogued Victrolas of that time could reach $500.00. *Courtesy of Alan H. Mueller.*

Right:
2-9. A 1922 Sonora catalogue exclusively devoted to posh period cabinetry. Although industrial carving machines could produce flawless, identical work, instruments of this sort could almost be considered "hand made" since they sold in such limited numbers. The retail prices were high (often $500.00 or more) and labor was comparatively inexpensive, so the larger phonograph companies all tried their hands at the elite cabinet game.

2-10. The 1919 Columbia Grafonola "Small Gothic" ("P-26") was one of six Gothic styles of "Period Grafonolas." Originally priced at a hefty $675.00, the "P-26" featured a heavily carved oak cabinet with muted polychrome highlights. *Courtesy of Brice D. Paris.* (Value code: VR)

2-11. A close-up of the "Small Gothic" Grafonola shows the detail which commanded such a high price tag. The right-hand door served as the grille for the horn. Despite its impressive appearance and carvings, the "P-26" was sold off at $450.00 during the recession of 1921. *Courtesy of Brice D. Paris.*

2-12. A 1921 Columbia advertisement for its "Period Design" Grafonolas. Illustrated was the "Charles II," but the Gothic models were similar in their vertical orientation and heavily carved panels. *Courtesy of Alan H. Mueller.*

2-13. Another "Period Design" Grafonola advertisement. *Courtesy of Alan H. Mueller.*

2-14. The Lampograph Company of St. Louis first advertised its unusual wares in the December 1918 issue of the *Talking Machine World*. This 66" high floor lamp carries patent dates of April 15, 1919 and July 22, 1919. Unlike other lamp phonographs of the period, many of which were table models, this model might have required a petite housewife to mount a stepstool in order to change a record! *Courtesy of the Domenic DiBernardo collection.* (Value code: D)

Right:
2-15. The "Capitol Phono-Lamp," was made by the Burns-Pollock Electrical Manufacturing Company, Indiana Harbor, Indiana. The obvious similarity between this device and the Lampograph in the previous illustration suggests a relationship between the two companies. *Courtesy of the Domenic DiBernardo collection.* (Value code: D)

2-16. With its shade opened, the "Capitol Phono-Lamp" (Style "OJR") was ready to play. Sound was projected downward through the base. 27" overall height. *Courtesy of the Domenic DiBernardo collection.*

2-17. This unmarked phonograph from the early 1920s was designed to resemble an urn. The sound was conveyed into the base, from which it exited through the wooden fretwork. With the lid in place and the crank positioned toward the wall, the disguise would be complete. *Courtesy of Pat and Jamie Jones.* (Value code: VR)

2-18. The Endlessgraph Manufacturing Company of Chicago produced this massive, 39" high lamp phonograph which was called, strangely enough, the "Fairy." *Courtesy of the Domenic DiBernardo collection.* (Value code: D)

2-19. With its lid removed, the playing mechanism of the "Fairy" lamp phonograph could be seen. The sound traveled down the right-hand arm and out through the grille in the base. *Courtesy of the Domenic DiBernardo collection*.

2-20. The underside of the lid displays the light blue "Fairy" decal. Note the wood-grained metal base. "Fairy" lamps were sold in a variety of finishes, including oxidized bronze and verdigris. *Courtesy of the Domenic DiBernardo collection*.

2-22. Following the Victor convention of relieving the starkness of its black oriental finishes with an interior pastel shade of red, blue or green, the interior surfaces of the "IV" shines in bright contrast. *Courtesy of John P. Andolina, Jr. "The Early Soundman."*

2-21. We might naturally imagine painted oriental finishes to be the exclusive domain of the larger, more expensive talking machines. Not so, as this Victrola "IV" attests. The "IV" was Victor's least expensive internal horn model for a number of years. Yet, here it has been spectacularly adorned. *Courtesy of John P. Andolina, Jr. "The Early Soundman."* (Value code: VR)

2-23. Even the otherwise utilitarian tone arm support bracket of the "IV" has been made elegant. *Courtesy of John P. Andolina, Jr. "The Early Soundman."*

2-24. A delicate fan graces the area around the speed control. The ordinarily modest molding is crowded with embellishment. Because of the diminutive size of the "IV," the density of decoration has far surpassed that of similarly finished floor models. *Courtesy of John P. Andolina, Jr. "The Early Soundman."*

2-25. The Modernola was more a "phonograph-lamp" than a "lamp-phonograph." The formidable dimensions of its cabinet reduced the lamp to a mere adjunct. The Modernola Company of Johnstown, Pennsylvania, began advertising in the *Talking Machine World* in January 1920, although the trademark had been used since 1918. *Courtesy of Walt and Ellen Bellm.* (Value code: VR)

2-26. The Modernola Company, like virtually every other phonograph manufacturer in the late 'teens and early 1920s, focused on cabinetry. This particular example employed an electric motor. *Courtesy of Walt and Ellen Bellm.*

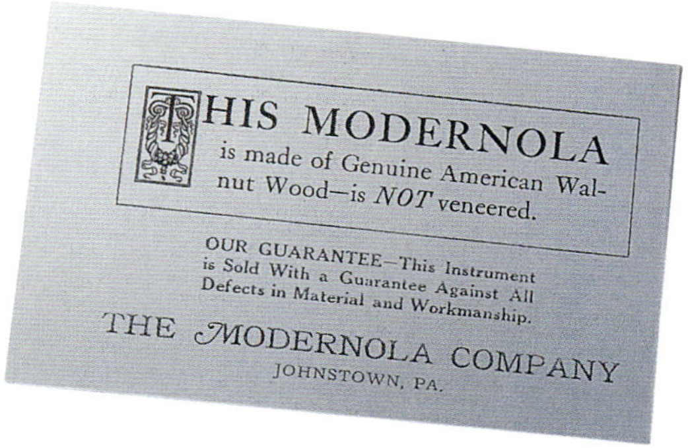

2-27. The Modernola Company assured its customers that its product (which meant the cabinet since the mechanical parts were generic) was one of quality.

2-28. The hand-painted images on the front doors lent this Modernola an "Audubon" ambiance. *Courtesy of Walt and Ellen Bellm.*

2-29. Another panel evokes Renaissance art despite its charmingly clumsy execution. *Courtesy of Walt and Ellen Bellm.*

2-30. In the triangular realm, the Davis phonograph was a distinguished inhabitant. The Davis Phonograph Company of Chicago offered several "corner models" (as well as conventional uprights and table models) during the post-WWI period. In this illustration, the obligatory Queen Anne legs, popular in early twentieth century America, and the filigreed grille are the only elements that disrupt the geometric harmony of the design. *Courtesy of Steve and Ellie Saccente.* (Value code: F)

2-31. Davis instruments sometimes incorporated wooden tone arms. The peculiar soundbox employed an adjustable metal ring to dampen the diaphragm, and thereby regulate volume. *Courtesy of Steve and Ellie Saccente.*

2-32. While the triangular cabinet aptly reflected the shape of the internal acoustic horn, it left no room for record storage. *Courtesy of Steve and Ellie Saccente.*

2-33. In this example, an optional cupboard has been attached to the low shelf, allowing dust-free space for 78s. *Courtesy of Steve and Ellie Saccente.*

3-34. With the discovery and opening of the tomb of King Tutankhamen in 1922-23, an interest in ancient Egyptian design ensued. This Piramida phonograph embodied the very essence of popular Egyptiana from its overall pyramidal shape to the "hieroglyphics" decorating its doors. *Courtesy of Steve and Ellie Saccente.* (Value code: VR)

2-35. The Piramida's interior surfaces were painted a rich red. 12" discs could be stored in the bottom of the record compartment, and 10" discs in the top. Note also the painted lotus blossoms at the foot and head of each corner post. *Courtesy of Steve and Ellie Saccente.*

2-36. Like many other overtly stylish phonographs of the time, the Piramida utilized a generic tone arm, soundbox and motor for its playing mechanism. The show-stopping cabinet was the major selling point. *Courtesy of Steve and Ellie Saccente.*

2-37. The lid of the Piramida displayed a representation of Osiris, god of vegetation, judge of the underworld, and king of the dead. A harpist was painted on the right-hand side. *Courtesy of Steve and Ellie Saccente.*

2-38. A side panel of the Piramida featured a Pharaoh being attended by a rather pudgy servant. *Courtesy of Steve and Ellie Saccente.*

2-39. Another panel showed a Westernized interpretation of Hathor, an ancient Egyptian goddess of dancing and music. *Courtesy of Steve and Ellie Saccente.*

2-40. The application of Egyptian décor was not limited to large cabinet model phonographs. This example of the German Nirona phonograph sported unusually simplified Egyptian-style motifs. *Courtesy of Gregg Cline.* (Value code: VR)

2-41. Portable phonographs that became popular in the early 1920s rapidly became more sophisticated. Shown is an unmarked portable which resembled a lady's hatbox of 11 1/2" diameter. The crank escutcheon was the only clue. *Courtesy of Norm and Janyne Smith*. (Value code: G)

Below:
2-42. Opening the "hatbox," one discovered a phonograph packed away. *Courtesy of Norm and Janyne Smith.*

2-43. A few seconds' assembly resulted in a complete phonograph waiting to entertain at a picnic, on a train or at the seaside. The hatbox motif may have eliminated the clumsy look of a suitcase portable, but this instrument would have looked pretty silly being toted by a gentleman to a day of golf. *Courtesy of Norm and Janyne Smith*.

2-44. What appeared to be a well-built secretary with matched veneers and Queen Anne legs was actually a phonograph cabinet: the Pathé "Classique" (as evidenced by the crank at right). Note the cruciform inlay on the lower doors. After the First World War, Colonial Revival became a popular furniture style in America, and this handsome cabinet would have complimented such a décor. *Courtesy of Freddie and Jeanette Blair.* (Value code: F)

Right:
2-45. The interior of this imposing cabinet contains an equally imposing mechanism: the Pathé "Actuelle," utilizing a paper cone diaphragm. Pathé announced this innovative system in the November 1918 issue of *Talking Machine World*. Company president E.A. Widmann was quoted as saying, "The object of this device… is particularly directed to the attainment of direct propagation in free air from a record of self-sustaining sound waves, substantially corresponding to the original recorded sounds, in intensity or amplitude, as well as in pitch and timbre." (…And if you believe that, we have some Florida swampland we'd like to sell you.) It is readily apparent in this view that much of the cabinet's height was purely aesthetic. Measurements are 28 3/4" wide x 20" deep x 64" high. *Courtesy of Freddie and Jeanette Blair.*

2-46. When phonographs masqueraded as articles of furniture, any function other than the reproduction of music was often curtailed or negated. The Grafonola "Regent" actually could not be used as a desk (see caption 1-44). The Aeolian "900-2" really could not be employed as a table (see caption 1-46). This mahogany dining room buffet, circa 1920, in Queen Anne style of exaggerated proportions (60" wide, 25" deep, 60" high) was a functional piece of furniture. The capacious drawers could hold linens and silverware. The broad upper surface could support serving pieces. The mirror was large and gracefully arched. However, this attractive buffet was also a phonograph. *Courtesy of John W. Hess.* (Value code: VR)

2-47. The right side of the buffet opens to reveal a substantially-built talking machine. Note the highly unusual die-cast tone arm with its bends reminiscent of a band instrument. The soundbox is a "Gilbert Tone Reflector," but there is no other identifying information marking this instrument. On the left side of the piece is record storage. Positioning the winding crank behind a door meant there was no telltale escutcheon on the exterior of the cabinet to shatter the impression of an ordinary dining room sideboard. *Courtesy of John. W. Hess.*

2-48. The Victrola "260" was introduced in 1922 for $160.00. It was one of the so-called "humpback" models which resulted from Victor's reluctance to abandon its anachronistic domed lid. These models were not popular. In 1923, the price of the "260" was reduced to $150.00, where it remained until the model was discontinued in 1925. This example was delivered with a special oriental finish, probably the "Japanese Lacquer" announced by Victor in December 1919. *Courtesy of the Domenic DiBernardo collection.* (Value code: F)

2-49. The interior of this Victrola "260" displayed the pastel treatment of Victor's oriental finishes. Note the unusually ornate embellishment of the "Victrola" decal on the underside of the lid. With the craze for mah jongg and the popularity of orientalia in the early-to-mid 1920s, such finishes may have helped boost sluggish sales of unpopular models. *Courtesy of the Domenic DiBernardo collection.*

2-50. Not all design elements in a phonograph were made solely for aesthetic reasons. The Victor affiliate in England marketed a table model phonograph (HMV Model "460," 1924) which utilized the delicate yet highly efficient Lumière diaphragm. This unusual acoustic "speaker" had been patented by Louis Lumière some 15 years prior to being put into production. The company chose to distinguish the rather plain cabinet with patterned quartered oak. *Courtesy of the Domenic DiBernardo collection.* (Value code: E)

2-51. Opening the lid revealed the gorgeous contrast between the wood grain, gold-plated metal parts, and the pleated, gilt Lumière diaphragm, which ingenuously suggested an Art Deco sunrise. *Courtesy of the Domenic DiBernardo collection.*

2-52. A close-up of the interior of the HMV shows the subtle yet deft use of color by its designers. *Courtesy of the Domenic DiBernardo collection.*

2-53. In playing position, the Lumière-equipped HMV more resembled a piece of sculpture than a phonograph. The fact that every component was designed for utility made its almost-accidental beauty all the more remarkable. *Courtesy of the Domenic DiBernardo collection.*

2-55. A 1927 Victor booklet entitled “A Musical Galaxy.” The colorful cover may be considered “pre-Deco.” The artist showed strong influences of “modern art,” especially Cubism. Active, iconoclastic designs such as this would soon be tamed by the rigid, static forms of American Art Deco. *Courtesy of Lou Caruso.*

2-54. In the United States, the period of design that immediately preceded the flowering of American Art Deco (after 1930) was a hodge-podge of “modern art” influences. In this 1927 Victor advertisement, female figures who seem to have stepped out of a Symbolist painting are framed by a border that could only be described as “Pennsylvania Dutch!” *Courtesy of Alan H. Mueller.*

2-56. A record duster from the formidably-named “M.W. Stevens' Estate” advertising Brunswick Panatropes. *Courtesy of Walt and Carol Myers.*

2-57. The massive and expensive ($1,100 less tubes) Edison "C-1" was introduced in September 1928. The cabinet of this combination radio-phonograph was fashioned from walnut in the style of an Italian credenza (U.S. Design Patent No.79,552 awarded to Noel S. Dunbar of New York City). *Courtesy of Chuck Azzalina.* (Value code: VR)

2-58. The soundbox of the Edison "C-1," designed by Theodore M. Edison (son of the great inventor), was capable of playing both lateral and vertical cut records. Unlike many other phonographs, the "C-1" radiated style even in the opened position. *Courtesy of Chuck Azzalina.*

2-59. If bigger hats should become popular, why not bigger hatboxes? This model looked even more like luggage than did "suitcase" portables. *Courtesy of Walt and Carol Myers.* (Value code: G)

2-60. This large "hatbox" phonograph was marked: "Special design made exclusively for James K. Polk, Inc., Atlanta, Georgia, Richmond, Virginia." The interior conveyed a distinctively feminine look. *Courtesy of Walt and Carol Myers.*

2-61. What appears to be a hatbox or vanity set with a padded, embossed lid is actually a prototype 1929 electric phonograph from the vaults of Edison's factory in West Orange, New Jersey. *Courtesy of the Edison National Historic Site*. (Value code: VR)

2-62. A neat appearance with definite Art Deco influences is apparent in this open view. This phonograph was designed to be plugged in and played through a radio amplifier. The volume control is located on top of the tone arm support. The electric motor was made by the Allen Hough Rotor Corporation of Milwaukee, Wisconsin. This phonograph was one of only a small number of models envisioned by Edison to play lateral-cut discs. This prototype never reached production. *Courtesy of the Edison National Historic Site.*

2-63. *Courtesy of Nina and Dave Heitz.*

2-64. The Mannerist attenuation of the suave figures in this 1929 RCA Victor advertisement for Radiolas (in this instance radio-phonographs) would not look out of place in an *haute couture* layout from the 1950s. *Courtesy of Alan H. Mueller.*

2-65. *Courtesy of Greg Bogantz.*

Chapter Three, 1930–1940: Reverberations of Deco

3-1. A German needle tin. The emergence of American Art Deco (after 1930) was directly influenced by European imagery, some of which had already been borrowed from the United States. In this way, the "hot jazz" icon of this tuxedoed saxophonist could become a universally recognizable evocation of modernity.

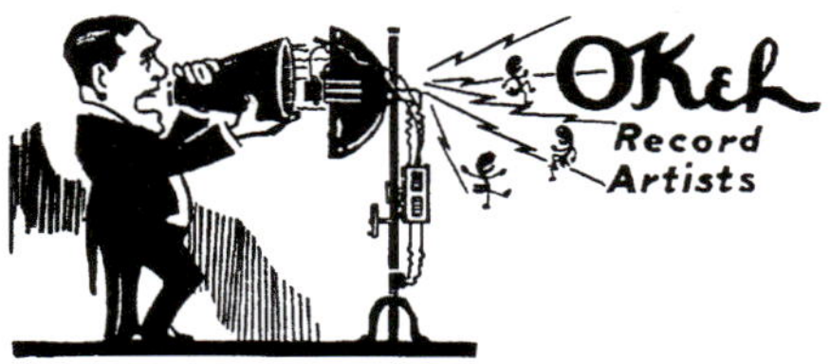

3-2. The 1930 Philco "Concert Grand" was an unusually contoured combination radio-phonograph: it fit into a corner. Its attractive cabinet emulated the "Italian credenzas" of the style period just ending. Priced high and introduced at the beginning of the Great Depression (not to mention its uncommon shape), production of the "Concert Grand" was limited to a few hundred. *Courtesy of Chuck Azzalina.* (Value code: E)

Right:
3-3. The opened position revealed the radio-phonograph controls and a stylish tapestry grille. *Courtesy of Chuck Azzalina.*

3-4. In the Philco "Concert Grand" record storage was cleverly concealed in the corner columns, a necessity of the wedge-shaped cabinet. Note the beautifully angled veneers and the emergence of American Art Deco elements in the speaker cloth. *Courtesy of Chuck Azzalina.*

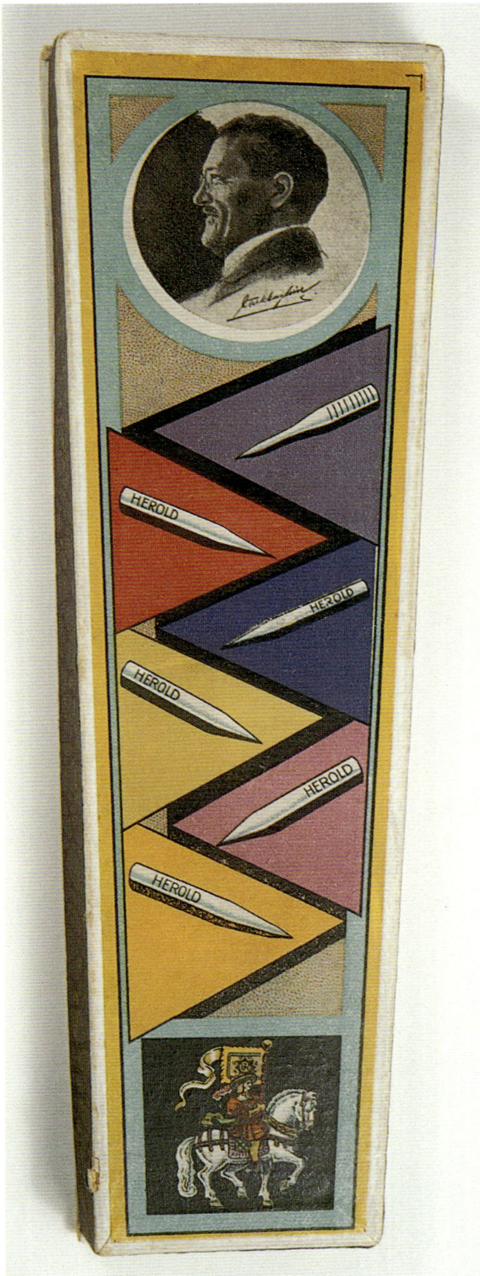

3-5. A German Herold needle box, 8" long. *Courtesy of Wilfried Sator*.

3-6. Three needle tins with a decidedly Art Deco flavor. *Courtesy of Walt and Carol Myers*.

3-7. Another trio of stylish European needle tins. *Courtesy of Walt and Carol Myers*.

3-8. Unique in its combination of "portability" and questionable taste was this 1930 Cecilian phonograph from Montgomery-Ward. At first glance, the cabinet appeared to be an old sea chest... (Value code: VR)

Right:
3-9. Opening the "chest" revealed a phonograph mechanism. The 1930 Montgomery-Ward catalogue, No.112, described the phonograph this way: "Reproduction of an ancient treasure chest from 'Ye Olde Days When Ye Pirates Bolde Sailed Ye Spanish Seas,' live again in this cabinet covered in hand-tooled effect." The material covering the chest was described as "Spanish Morocco grained DuPont Fabrikoid."

3-10. Close-up the Cecilian lid decoration. Difficult to take seriously, this Cecilian "treasure chest" phonograph must have been a rare glimmer of fun during the Great Depression. Ye Price of Ye Olde Pirate Chest was $41.95 postpaid to landlubbers.

3-11. For *extreme* portability, few phonographs could match the Swiss-made Mikiphone. Looking more like an oversized pocket watch than a talking machine, this device could be carried in a large pocket. 4 1/2" diameter. *Courtesy of Walt and Carol Myers.* (Value code: G)

3-12. The Mikiphone's ingenious arrangement included a plastic resonator mounted directly to its soundbox. The arm at left swung outward to allow 10" discs to be played; it also activated the motor. The watch-like winding key may be seen in the foreground. *Courtesy of Walt and Carol Myers.*

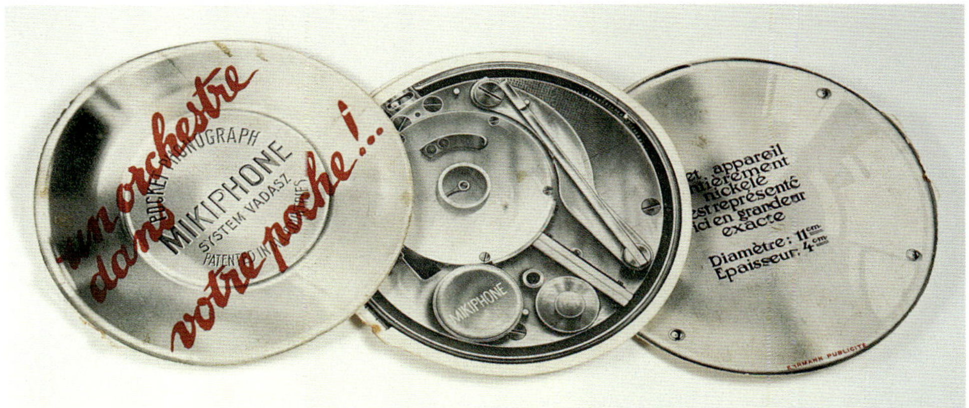

3-13. A Mikiphone brochure printed in French. The Mikiphone was very popular the world over during the late 1920s and 1930s. *Courtesy of Arophon.*

3-14. The Odeon company of Germany manufactured this charming red 17 3/4" diameter portable phonograph in a textured metal case. *Courtesy of Steve and Ellie Saccente*. (Value code: G)

3-15. Like the Mikiphone, this Odeon employed "System Vadasz," which referred to patented reproducing features. *Courtesy of Steve and Ellie Saccente.*

3-16. Once assembled, however, the Odeon displayed an ostentatious tone arm that snaked from within the machine's limited dimensions. *Courtesy of Steve and Ellie Saccente.*

3-17. This European-made Coronet was typical of the modestly-scaled phonographs which became available following a portable craze during the late 1920s. As usual, the parts folded up compactly. However, in this case the unexpected purple of the turntable felt and horn interior lent considerable panache. *Courtesy of Waves.* (Value code: H)

3-18. The name Capehart would enjoy decades of recognition in the phonograph industry. At first, the brand was associated with coin-operated automatic phonographs (or juke boxes). Later, Capehart would become a prestige name in home record changers. The formation of the Capehart Automatic Phonograph Company was announced in the March 1928 issue of *Talking Machine World*. The firm produced a juke box phonograph called the Orchestrope. In March 1930, Capehart announced its Amperion automatic phonograph (shown), a lower-priced model. The Amperion was produced over the next few years, after which the company aimed its sights at the home market. The cabinet of this machine has vague elements of Jacobean styling, most notably the bulbous feet. *Courtesy of Nina and Dave Heitz.* (Value code: E)

3-19. The record changer of the Amperion Model "100" sported the "crinkle" paint which would become a popular finish for phonograph mechanisms during the "Industrial Modern" period of the late 1930s and 1940s. In this instrument, circa 1930, the crinkle effect is all the more striking for its uncommonly bright color. *Courtesy of Nina and Dave Heitz.*

3-20. RCA's "RAE-84" radio-phonograph of 1932 may look like a refugee from the set of Universal's *Frankenstein*, but it incorporated up-to-date technology within its heavy styling (in what could only be described as Jacobean Gothic). RCA described the "Bi-Acoustic" system built into this and other instruments as follows: "The new 'Bi-Acoustic' Circuit, however, is provided with a series of Tone Equalizers. These equalizers are hollow boxes with various sized openings like the pipes of an organ, and are capable of being 'tuned' to the particular cabinet resonances which cause tone distortion. As a result, the 'Bi-Acoustic' Circuit eradicates cabinet booming..." Another innovation was the ability to play 33 1/3 speed discs that RCA called "Long-Playing Records" or "Program Transcriptions." Furthermore, this instrument was equipped with a home recording outfit. *Courtesy of Greg Bogantz.* (Value code: G)

3-21. The playing compartment of the "RAE-84" included, on the left, a formidable record changer in then-popular crinkle finish. Behind the turntable, a meter showed the input level for home recording. To the right, the "modern" styling of the RCA tone arm is a sharp contrast to the gloomy medievalism of the cabinet. *Courtesy of Greg Bogantz.*

3-22. An RCA Victor cardboard store display, overall width 42 1/2" x overall height 31", promoting the home recording feature. Note the small hole under the right hand of the mother, through which a wire was once threaded and attached, for realism, to an instrument such as the "RAE-84." *Courtesy of Lou Caruso.*

3-23. A dealer's display promoting a home voice recording as "A Modern Christmas Card." 42 1/2" overall width, 31" overall height. *Courtesy of Lou Caruso.*

3-24. The artful record sleeve of a "Long-Playing Record" from RCA Victor's short-lived 33 1/3 rpm experiment. Although the advantages of long playing records would be universally appreciated some 16 years later, the Great Depression proved an inopportune time to promote revolutionary technology. A 1932 Victor brochure optimistically proclaimed, "Now! Modern Home Entertainment..." *Courtesy of Greg Bogantz.*

3-26. A record-cleaning brush, measuring 3" diameter, with a tropical scene lush enough for Dorothy Lamour. *Courtesy of Wilfried Sator.*

3-25. An unusually stylish HMV record duster. *Courtesy of Walt and Carol Myers.*

3-27. A glass tip tray advertising Burchard needles. 6" long. *Courtesy of Wilfried Sator.*

3-28. A jointed wooden figure advertising RCA "Radiotron" tubes. 15 1/2" high. *Courtesy of Fred Guenther*.

3-29. An 8" diameter RCA advertising fan of the period. *Courtesy of Fred Guenther.*

3-31. Opening the center doors of the "D-22" allowed access to the radio-phonograph controls. This was RCA's top-of-the-line model at the time. *Courtesy of Daniel Gustafson.*

3-30. The 1935 RCA "D-22." Despite its almost square shape, this imposing instrument made delicate use of the artful veneering associated with Deco. *Courtesy of Daniel Gustafson*. (Value code: F)

Below:
3-32. The design of the "D-22" record changer was tiered, sculpted. With their stylized musculature, the changer arms were virtually alive with the optimistic pulse of progress. The mechanism played a stack of 78s from the top down, literally slinging a record off the pile after it had been played. *Courtesy of Daniel Gustafson*.

3-33. A German display for sapphire needles. The individual plastic containers were stylish enough to be ladies' compacts. *Courtesy of Wilfried Sator.*

3-34. A German sign, measuring 5" x 7", advertising rather peculiarly-shaped phonograph needles. *Courtesy of Wilfried Sator.*

3-35. Display pieces, in German, for Cona needles. Placard measures 6" x 9". *Courtesy of Wilfried Sator.*

3-36. A store placard, in Spanish, advertising phonograph needles, measuring 8" x 10". *Courtesy of Wilfried Sator.*

3-37. Replace the round radio dial of this 1935 E.H. Scott "Philharmonic" radio-phonograph in the "Roslyn Grande" cabinet with a TV screen, and you can recognize the visual evolution of early floor-model television sets. *Courtesy of Chuck Azzalina.* (Value code: E)

3-38. The lid of the "Roslyn Grande" cabinet, a hodge-podge of period furniture elements, lifted to reveal the phonograph. The interior components already reflected the "Industrial Modern" styling, in drab brown hues, which would peak in the 1940s. *Courtesy of Chuck Azzalina.*

3-39. The most remarkable aspect of this Scott radio-phonograph was the electric amplifier, shockingly chromed (pun intended). Many companies would rather not have drawn attention to the inner workings of their machines. However, E.H. Scott lent to its massive electronics the mirrored brilliance of a French palace. *Courtesy of Chuck Azzalina.*

3-40. The clean styling of the 1936 E.H. Scott "Waverly Grande" cabinet incorporated a Deco front panel, contrasting more traditional fluted corner pillars laid obliquely to draw the eye toward the center of the instrument. *Courtesy of Greg Bogantz.* (Value code: D)

3-41. The placement of the "Imperial Allwave 23" radio controls, deliberate yet graceful, emphasized this instrument's identity as top-of-the-line sound equipment. *Courtesy of Greg Bogantz.*

3-42. The record-playing mechanism was housed beneath a lid, which was well concealed since it fit flush against the notched molding of the upper cabinet. The corner thumb-holds, a necessary functional element, were subtly integrated into the cabinet design. *Courtesy of Greg Bogantz.*

3-43. The 78 record changer (Garrard "RC-10") installed in this Scott instrument maintained the customary brown color-scheme of the period, and no attempt was made to depart from the changer's practical aesthetics. Scott pioneered the "component" concept in sound equipment, and cabinets such as the "Waverly Grande" could be used to enclose various combinations of high-end technology. This instrument even contains woofers and tweeters, those specialized speakers commonly associated with more recent audio systems. *Courtesy of Greg Bogantz.*

3-44. A Bakelite Capitol needle dispenser from the late 1940s, measuring 3 1/2" high. The dispenser itself was designed to suggest the lines of a floor model phonograph from the period. The central control knob ejected needles into the scooped area below. Used needles were deposited at the top left. *Courtesy of David Werchen.*

3-46. Opening the central doors revealed a control panel more suitable for an aircraft than a home appliance. *Courtesy of Chuck Azzalina.*

3-45. The 1938 RCA "V-109" was housed in this handsome cabinet with the clean lines of "Industrial Modern" design. *Courtesy of Chuck Azzalina.* (Value code: F)

3-47. The record changer, although technically similar to the one shown in illustration 3-32, conveyed an appearance more purposeful than decorative. The utilitarian "crinkle finish" and severe lines were relieved only by the changer arm's suggestion of a streamlined locomotive. *Courtesy of Chuck Azzalina.*

3-48. A 1938 RCA advertising sign celebrating "Forty Years of Leadership." 28 1/8" x 22 3/4". *Courtesy of Robin and Joan Rolfs*.

3-49. Of all the portable phonographs ever built, perhaps none embodied more flair than the RCA Victor "Specials." Shown is a 1938-1939 "Special" in green with aluminum metal parts. *Courtesy of Daniel Gustafson*. (Value code: F)

3-50. Another RCA Victor "Special" in closed position. The "Specials" were not regularly catalogued by RCA, and it has been suggested that these electric portables were created specifically in honor of the 1939 New York World's Fair. The "Specials" emerged from the hand of John Vassos, a highly-acclaimed contemporary designer. *Courtesy of Daniel Gustafson*. (Value code: D)

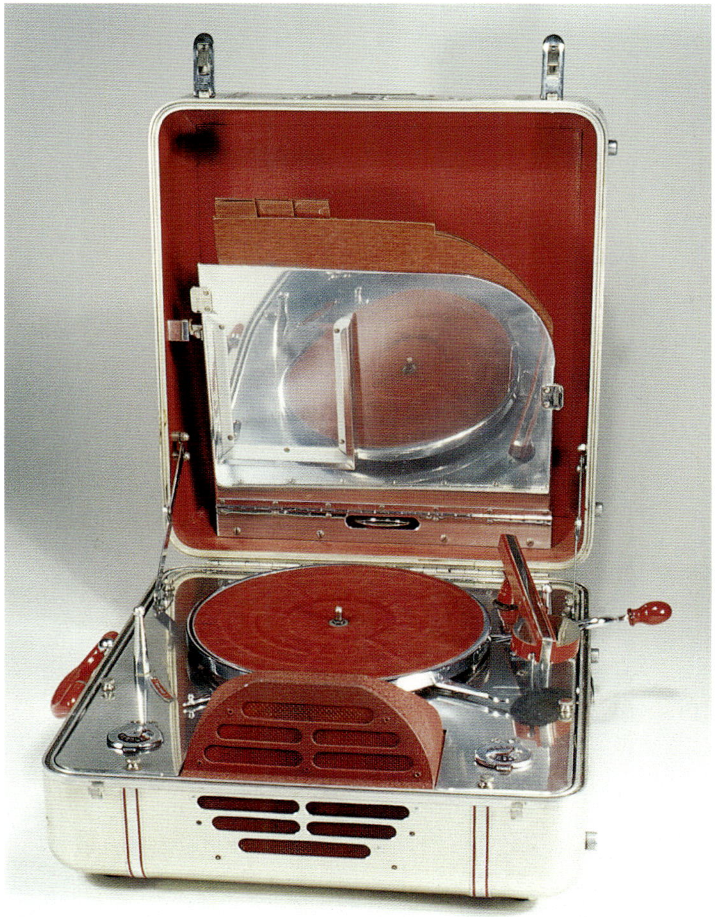

3-51. The previous example opened. How much more voguish could a portable phonograph be? *Courtesy of Daniel Gustafson.*

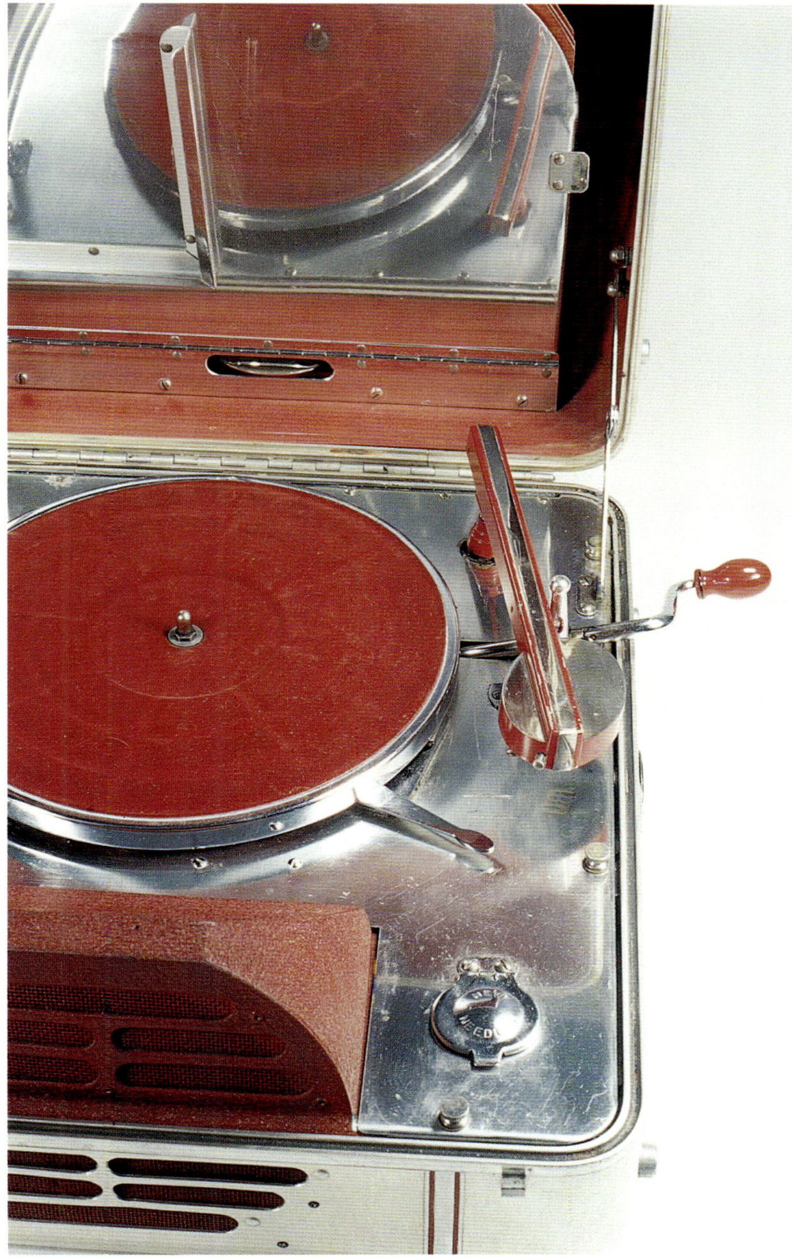

3-52. A close-up of the 1938-1939 red RCA Victor "Special." *Courtesy of Daniel Gustafson.*

Chapter Four, 1941–1949: Style in Transition

"From a Bluebird Record Sleeve"

4-1. In the late 1930s, and during the 1940s, a class of inexpensive table model phonograph arose which consisted of a turntable surmounted by a resonator dish. Although the record revolved electrically, the reproduction was strictly acoustic, a latter-day outgrowth of the "diaphragm" playing systems of earlier phonographs such as the Pathé Actuelle. In this example, kitsch imitates life, as a garish pressed paper sombrero amplifies the music. *Courtesy of Steve and Ellie Saccente.* (Value code: H)

4-2. A wave of pan-Americanism swept the United States in the late 1930s. Movies, music, art and fashion all reflected an interest in Mexican and Latin American motifs. However, this bizarre record player may be taking the Mexican Hat Dance to the extreme! *Courtesy of Steve and Ellie Saccente.*

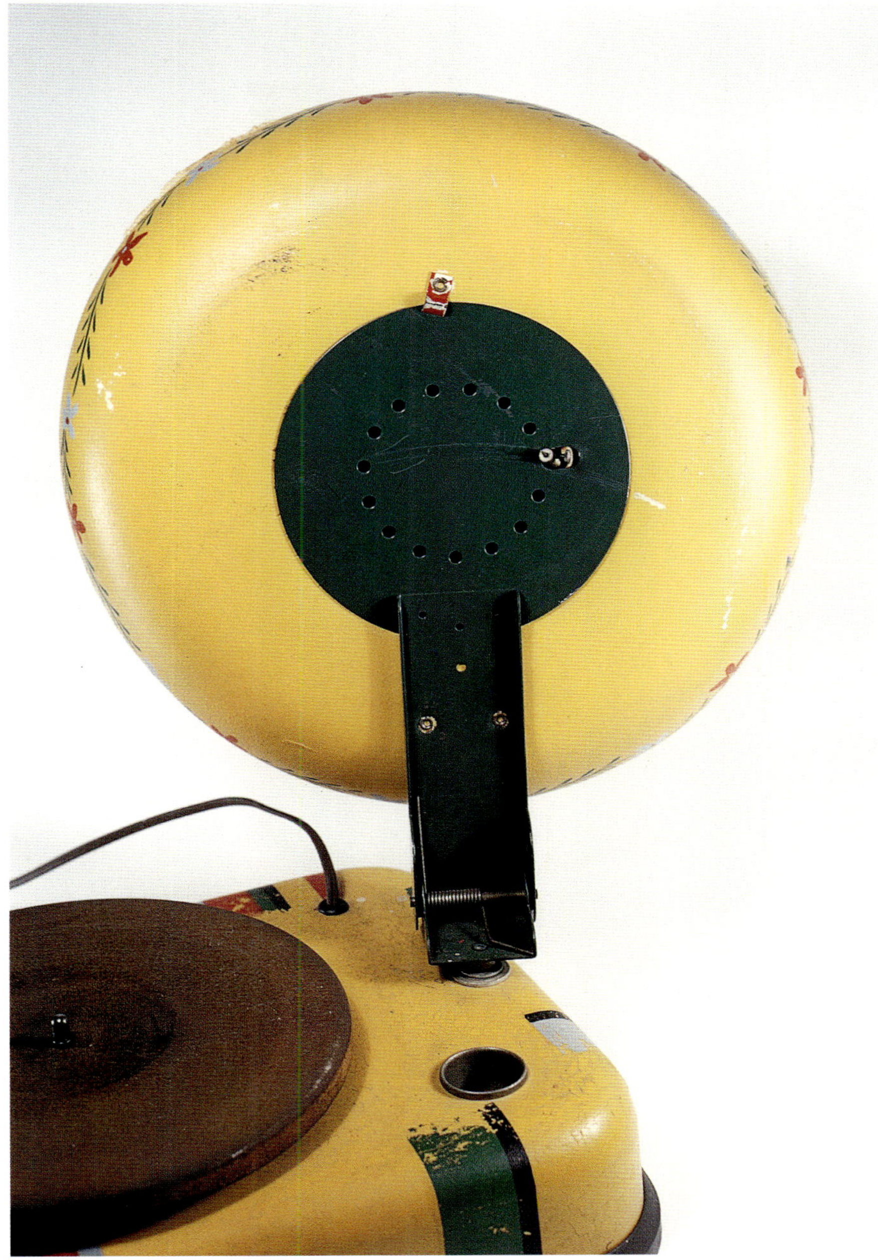

4-3. The workings of the simple resonator may be seen here. A needle chuck located in the three o'clock position held conventional steel needles, retrieved from the needle cup below. *Courtesy of Steve and Ellie Saccente.*

4-4. The cabinet of this inexpensive Emerson radio-phonograph, circa 1940, may have been plain, but the decorative frieze of the front panel was undeniably charming. Furthermore, the designer managed to preserve economy of scale by including a slot in the rear panel of the cabinet through which the record protruded as it played. *Courtesy of John A. Reinicke.* (Value code: I)

4-5. Looking like it was awaiting the masterful touch of Dick Tracy, the Philco "Mystery Control" (1939-1942) was the great-granddaddy of the remote controls which operate today's electronic equipment. When used in conjunction with a Philco radio-phonograph, the telephone-inspired dial (made of Plaskon, an early plastic) could call up preset radio stations, adjust the volume or activate the phonograph. A transmitter in the "Mystery Control" sent wireless signals to trigger these automatic functions in the radio-phonograph unit. *Courtesy of Greg Bogantz.*

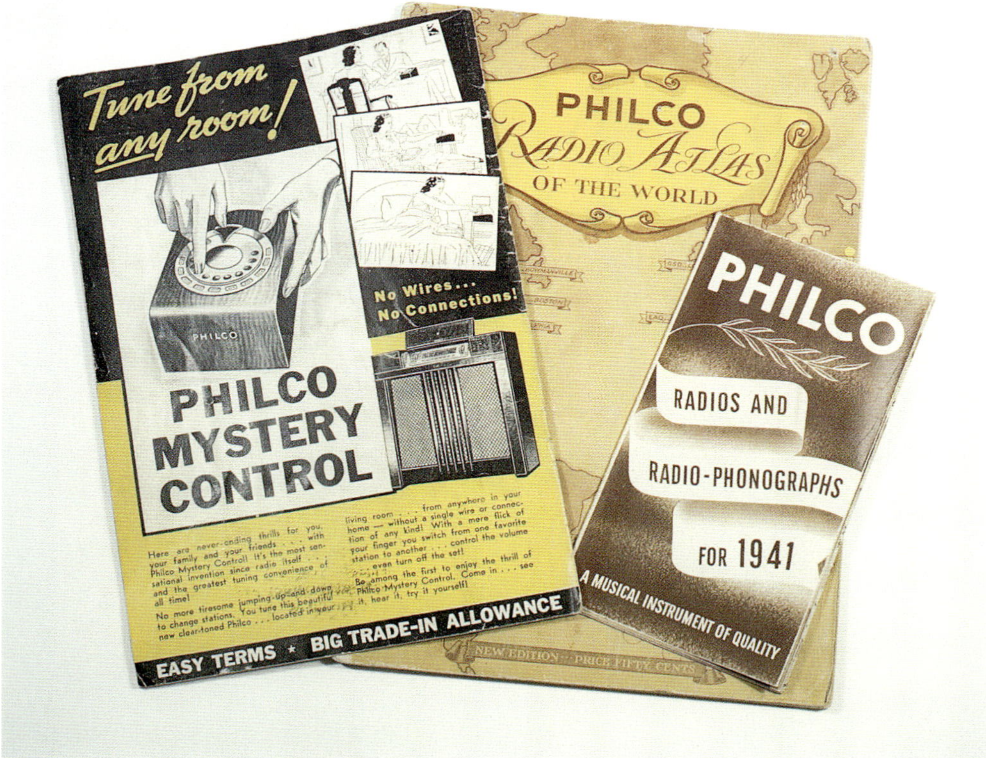

4-6. Promotional material for the Philco "Mystery Control" described the "...never-ending thrills..." that this marvel of modern science could provide. *Courtesy of Greg Bogantz.*

4-7. An illuminated Philco needle display piece, measuring 11" high. *Courtesy of Pat and Jamie Jones.*

4-8. An RCA neon advertising clock, measuring 21" in diameter, made by the Canadian Neon Ray Clock Company, Limited, Montréal and Québec. Model "7NR." *Courtesy of the Domenic DiBernardo collection.*

4-9. The RCA Victrola "V-135," circa 1940, had the unmistakably veneered look and rounded edges of Deco. However, the matter-of-fact arrangement of the controls, and the drab color-scheme, betokened the "Industrial Modern" influence. *Courtesy of Waves.* (Value code: I)

4-10. A gold plastic Philco dealer's plaque, 23" in diameter. *Courtesy of Waves.*

4-11. The Philco "42-1008" (1942) may have had the familiar slab sides, rounded corners and slatted grille which reeked of Deco styling, but its technical elements were truly innovative. *Courtesy of Greg Bogantz.* (Value code: H)

4-12. The most significant cabinet feature of the "42-1008" was its "tilt front." As a Philco catalogue stated, "No lid to lift... no need to remove decorations, no dark, hard-to-get-at compartments. Simply tilt forward the grille, place your records and tilt it back." The size, shape and light color of the pick-up were meant to focus attention on Philco's short-lived "Beam of Light" system of reproduction. As the catalogue announced, "The pointed, scraping needle is gone! Instead, a rounded jewel that never needs changing floats over the record and reflects the music on a beam of light from a tiny mirror to a photo-electric cell." *Courtesy of Greg Bogantz.*

4-13. A Philco catalogue from 1942. *Courtesy of Greg Bogantz.*

4-14. A 1940 Philco advertisement promoting "tilt-front" phonographs. *Courtesy of Greg Bogantz.*

4-15. An RCA advertisement from 1946. Of special note is the contest to award prizes for the oldest Victrola for each model year, from 1906 to 1925. It would be interesting to learn how many responses were received. *Courtesy of Greg Bogantz.*

4-16. This Philco "1201" radio-phonograph from 1946 is often called the "Bing Crosby." The singer was associated with Philco through broadcast sponsorship, and his picture frequently appeared in the advertisements for this instrument. Its design evoked the elevated control console of a streamlined train or airplane, with bright speaker cloth separating the radio (above) from the phonograph (below). *Courtesy of Dan and Sandy Krygier.* (Value code: I)

4-17. In the late 1940s, Philco pioneered a "no hands" record playing system. A 78 inserted through the hatch at the bottom played without human assistance when the door was closed. *Courtesy of Dan and Sandy Krygier.*

4-18. More traditional in appearance than the "1201" was the Philco "46-1203" (1946). The "sliced" look of the grille redeemed an otherwise bland cabinet. *Courtesy of Greg Bogantz.* (Value code: I)

4-19. A 1946 advertisement for the Philco "1203." *Courtesy of Greg Bogantz.*

4-20. Popular furniture of the 1940s with flowing, liquid lines was reflected in the 1947 RCA "650." The smooth, sculpted lid, rounded corners, and "headlamp" control knobs, virtually defined so-called "waterfall" styling of the period. *Courtesy of John A. Reinicke.* (Value code: I)

4-21. Mimicking an end table, this 1947 Farnsworth "GK267" "Chairside" would be inconspicuous in an up-to-date living room of the post-war period. The example shown is finished in walnut. *Courtesy of Greg Bogantz.* (Value code: I)

4-22. Conveniently accessible, as its name suggested, the "Chairside" housed a "P56" 78 record changer and an AM radio. Note the inverted radio dial, meant to be viewed while seated beside the instrument. A later generation would describe this kind of thoughtful design as "ergonomics." *Courtesy of Greg Bogantz.*

4-23. The blazing red pick-up of this Farnsworth "P-56-MP" 78 changer (1947) fairly rockets away from the slender arm. Can Mars be far away? Another vivid departure was the stark white of the flocked turntable. *Courtesy of Greg Bogantz.*

4-24. A 1947 advertisement touting the "GK267" "Chairside" and other Farnsworth instruments as "truly post-war." The blond finishes that would become so popular in the 1950s have already made their appearance. *Courtesy of Greg Bogantz.*

4-25. Another 1947 advertisement for Farnsworth instruments, some of which were offered in "genuine South American prima vera," a blond wood with stylishly rich grain. *Courtesy of Greg Bogantz.*

4-26. The most significant cabinet feature of this 1947 Capehart "111-N" was the use of vertical louvers over the speakers. Of course, these hinged panels were not intended to control volume, as they had in the pre-electric Columbia Grafonolas of the 'teens. *Courtesy of Greg Bogantz.* (Value code: G)

Below:
4-27. The Capehart "111-N" housed an AM/FM/short wave radio of straightforward styling. Note the FM dial is calibrated in the "old" FM standard (which was superceded in the early 1950s) showing "radio channel numbers" instead of 88 to 108 megahertz. *Courtesy of Greg Bogantz.*

4-28. The centerpiece of the Capehart was the company's complex 78 record changer (Model "41-E-MP"). The "amusement park" styling of this formidable mechanism required a cavernous playing compartment. The apparatus played both sides of the records with a continuous repeat. Shown are electrically-recorded discs from a period earlier than the instrument. *Courtesy of Greg Bogantz.*

4-29. Juxtaposed against the drab industrial finish of the changer, the bright, feathery arm and slender pick-up (with GE "variable reluctance" cartridge) lend sudden levity to the interior. *Courtesy of Greg Bogantz.*

4-30. A Capehart brochure and record duster with dealer name. *Courtesy of Greg Bogantz.*

4-31. This late-1940s Firestone radio-phonograph (with "Airchief" radio) is a peculiar amalgam of Deco and classical elements, which ultimately resembles the robots of the post-war imagination. The 78 phonograph slid out from the central, decoratively veneered panel. *Courtesy of Dave Barker.* (Value code: H)

4-32. It became a convention of the late 1940s to have a record changer that pulled out or was cantilevered out from the center of a radio-phonograph. Ergonomic planning, however, was not evidenced in the Firestone's cramped drawer that withdrew barely enough to allow access to the turntable. *Courtesy of Dave Barker.*

4-33. The striking impact of this Zenith "12H090" radio-phonograph arose from the visual textures of its decorative veneers. The style of the cabinet was essentially Empire, though (considering audio manufacturers' cavalier attitude toward periods of design) it might have been marketed by the company under almost any name. With its technical components hidden, this Zenith presents itself as an extraordinary piece of furniture. *Courtesy of Dave Barker.* (Value code: H)

4-34. Philco "tilt-front" be hanged – the Zenith "12H090" had a *double* tilt-front! The signature Cobra pick-up arm (complete with realistic "eyes") was present, along with the ubiquitous push buttons that betokened ease of operation. Push button convenience was a perception of "modernity" which would influence the design of home appliances from the 1940s through the 1960s. *Courtesy of Dave Barker.*

4-35. At the end of the 1940s, the "Bing Crosby Junior Juke" was a popular child's record player. A crystal pick-up played 78s through a small, tube amplifier. The curved plastic panels that bridged the front of the machine were illuminated from behind. The playing compartment was left open for convenience. Not exactly the spit and image of a Wurlitzer 1015, but sufficient to enliven any kid's clubhouse. *Courtesy of Dan and Sandy Krygier.* (Value code: H)

4-36. The Model "60 PR" Phonograph Radio Combination by the Ray Dyne Manufacturing Corporation, New York City. With minimal effort, it managed to suggest the look of a grownup floor-model phonograph, circa 1950. *Courtesy of Dan and Sandy Krygier.* (Value code: I)

4-37. 78s were played by ordinary steel needles, which persisted for use in children's phonographs long after adult equipment had converted to jewels. The imprecise calibrations of the radio dial were hopefully offset by the fun of tuning in to *Captain Midnight*. *Courtesy of Dan and Sandy Krygier.*

4-38. A 1947 Admiral radio-phonograph (cabinet No. "6C71," chassis No. "10A1"). The pneumatic look of the body suggested those capacious mohair armchairs which graced the front rooms of America during the 1930s. Despite this Deco holdover, a new era of post-war cabinet design was about to begin. *Courtesy of Dan and Sandy Krygier.* (Value code: H)

Right:
4-39. A 78 record changer emerged from the center of the Admiral, a design feature shared by other instruments of the period. *Courtesy of Dan and Sandy Krygier.*

4-40. A look at the Admiral's control panel suggests the lingering impression given by many radio-phonograph combinations of the 1940s: that the phonograph was an appendage to what was essentially a floor model radio. *Courtesy of Dan and Sandy Krygier.*

4-41. When television began its phenomenal rise to popularity after WWII, the radio-phonograph was already common in the American home. Although TV sets were often sold as stand-alone units, it was inevitable that some manufacturers would insert the new technology within the older radio-phonograph structure. This General Electric advertisement from the May 10, 1948 issue of *Life* illustrated two such instruments. When three home entertainment devices were housed within a single cabinet, a descriptive catch-phrase was coined which is still very much in use. Note, at the bottom left, the Model "901" was called a "de luxe Entertainment Center."

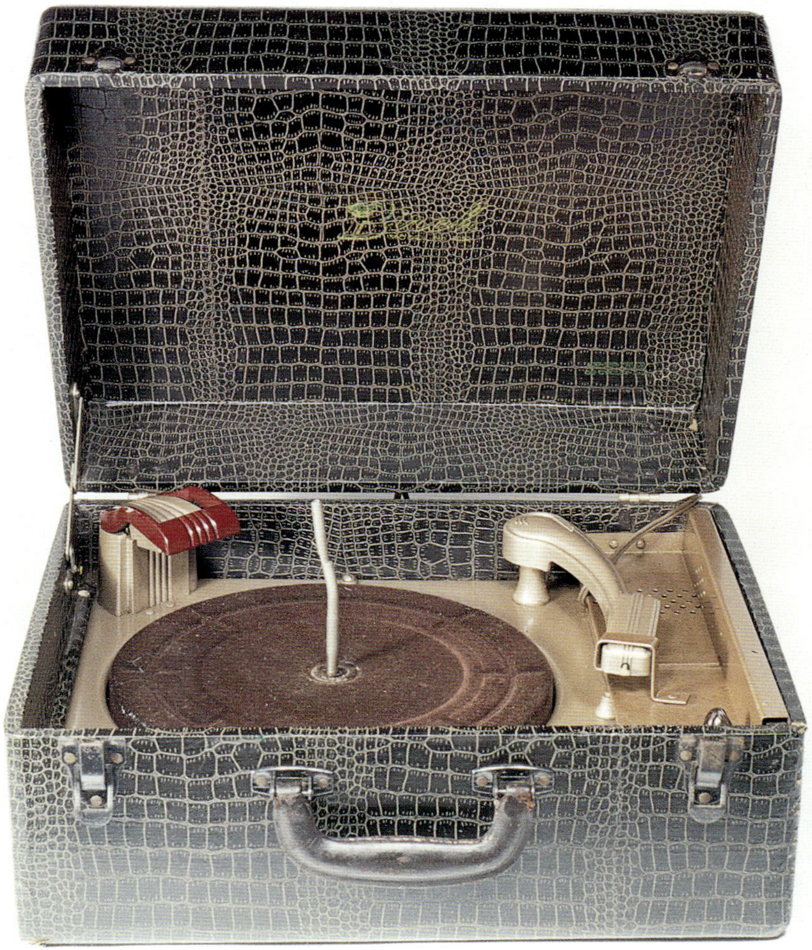

4.42. Unusual in faux-alligator is this post-war Birch plug-in portable. The one splash of color amid the earthy tones is a curious departure. *Courtesy of Dave Barker.* (Value code: I)

4-43. A characteristically 1940s Fidelitone needle display. *Courtesy of John P. Andolina Jr. "The Early Soundman."*

4-44. In the RCA "63E" the usual impact of "Industrial Modern" styling has been distilled to the melding of geometric forms: circle, rectangle and triangle. Nothing about the compact plastic body interferes with the flow of its lines – even the pick-up arm is retractable. *Courtesy of Greg Bogantz.* (Value code: I)

4-45. The "63E" pick-up arm popped up to reveal the volume control. This knob might otherwise have disrupted the symmetry of the whole, because of its offset placement. *Courtesy of Greg Bogantz.*

4-46. A Christmas advertisement featuring the "63E." *Courtesy of Greg Bogantz.*

4-47. Viewed from the front, the 1948 RCA "77U," could be mistaken for an ordinary table radio. The sinuous outline of the phonograph lid, however, foreshadowed the parabolic imagery that would pervade the 1950s, when swoops and accelerated curves disrupted otherwise static objects. *Courtesy of Dave Barker.* (Value code: I)

4-48. Although the playing compartment of the "77U" retained the basic brown of the pre-WWII period, the gold-colored components of the 78 changer presaged the bright plastic elements which would adorn record players during the following decade. *Courtesy of Dave Barker.*

4-49. The Model "1401" radio-phonograph was a late-1940s adaptation of Philco's "toaster" record playing system (see illustration 4-16). With an aeronautic sweep, the speaker grille emerges from the top of the cabinet. *Courtesy of Dan and Sandy Krygier.* (Value code: I)

4-50. The recorded format that signaled the death knell for the venerable 78 rpm record was the 33 1/3 rpm disc revived by Columbia in 1948, 16 years after RCA's failed experiment with the same speed. This 1948 Model "M101" helped launch the "LP" (Long Playing) record, which would dominate for the next four decades. Columbia did something earth-shattering, and did it with style. The appearance of this instrument was just as innovative as the new microgroove technology. The design was sleek and sophisticated, incorporating minimal profile within parallel lines. *Courtesy of Bob Havalack "Eager Bros. Music."* (Price code: I)

4-51. The debonair styling of the "M101" interior included only a turntable and pick-up arm. Like some of RCA's first 45 rpm players that were to follow, this instrument was meant to be plugged into an existing phonograph or radio. It functioned as an auxiliary unit to allow the new 33 1/3 rpm records to be enjoyed. *Courtesy of Bob Havalack "Eager Bros. Music."*

4-52. The front panel of this 1949 Zenith "6R980-LP" foreshadowed Studebaker styling of the 1950-51 period. The cabinet was otherwise unremarkable, but the interior held a surprise. *Courtesy of Greg Bogantz.* (Value code: I)

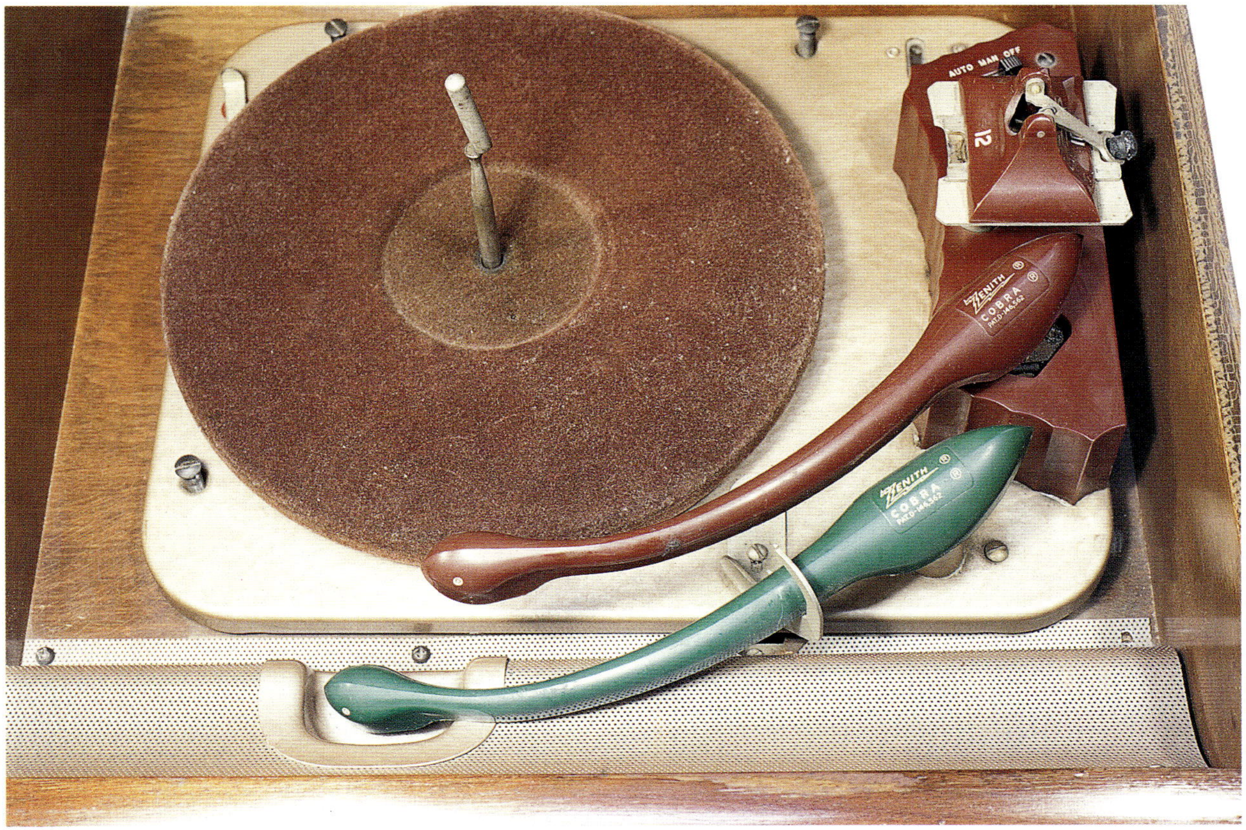

4-53. The reptilian fantasy of the playing compartment included not one but two of Zenith's signature "Cobra" pick-up arms. The red arm worked automatically with the changer mechanism to play 78s. The new 33 1/3 rpm LPs were played by the green arm, which was manually operated. *Courtesy of Greg Bogantz.*

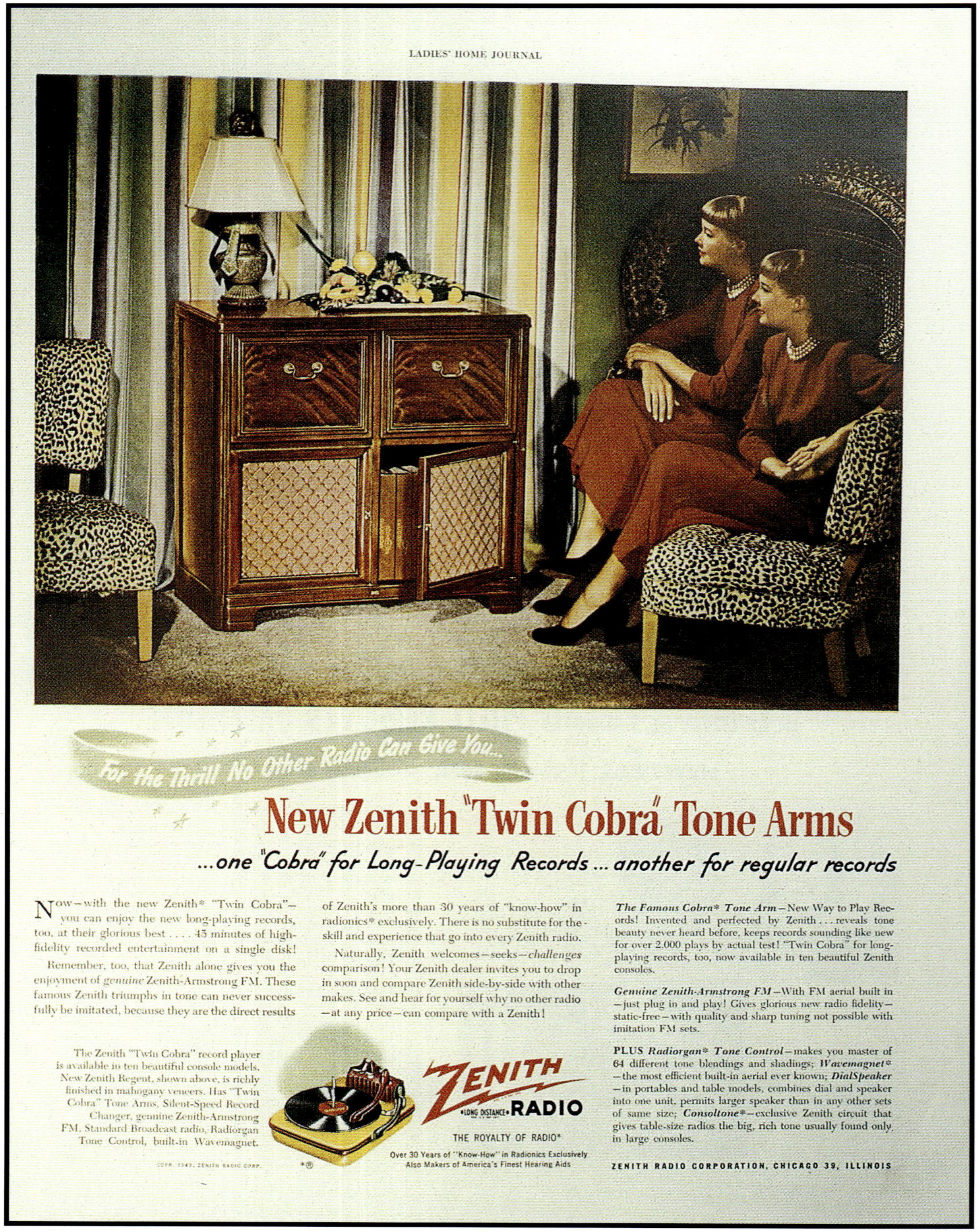

4-54. An advertisement for the Zenith "Twin Cobra" system. Although the Zenith dual arm technology was short-lived, advertisements for other products exploiting a cute set of twins would endure – witness the Toni Twins of hair care fame, and Doublemint's gum-chewing doppelgängers. *Courtesy of Greg Bogantz.*

4-55. In February 1949, RCA Victor introduced the (7" diameter) 45 rpm record with much hoopla, a move initially designed to produce greater value and convenience. Rival Columbia had inaugurated its own Long Playing discs the previous year, though it's ironic that Columbia chose 33 1/3 rpm (7", 10" and 12" sizes). RCA had briefly marketed a 33 1/3 rpm record in 1931-32. Why wouldn't the company return to this speed when it revived an alternative to the 78? RCA's 33 1/3 record of the early 1930s had met a precipitate death during the Depression. Once bitten, twice shy, perhaps. So, RCA and Columbia, with their competing formats, began slugging it out as they had at the turn of the century. This "9-EY-3" was RCA's first 45 rpm player with an amplifier and speaker. *Courtesy of Kurt Wilmarth "Echo-Tone Music."* (Value code: H)

4-56. One of RCA's first 45 changers was the 45J (1949), which was designed to plug into a contemporary radio such as the one shown. A 1950 RCA advertisement stated, "It can play through any radio, phonograph, or television set. Changes a stack of '45' records automatically from inside the big center spindle – no posts or clamps to adjust." The changer and radio evinced subdued elegance, except for the flashy red oilcloth covering of the turntable. *Courtesy of Greg Bogantz.* (Value code: I)

4-57. This 1949 Motorola "59F11" 45 player with AM radio looks like it could have doubled for a Geiger counter in a post-war Republic movie serial. The exterior was uninspiring, but the interior, with its lavish use of gold and bright plastic, was downright "rocket age." *Courtesy of Bob Havalack "Eager Bros. Music."* (Value code: H)

Below:
4-58. In the "war of the speeds" that occurred after 1949, this Zenith "Twin-Seven" was an early compromise. Capable of playing either 45 or 33 1/3 rpm discs (in 7" size only) with a single, short Cobra arm, this phonograph was the opposite approach of the "Twin Cobra" unit seen in illustration 4-53. The "Twin-Seven" was housed in a simple cabinet designed to be installed in a radio-phonograph, blending unobtrusively. The dual turntables and the pick-up arm which could pivot 180 degrees were the most noteworthy features. *Courtesy of Bob Havalack "Eager Bros. Music."* (Value code: VR)

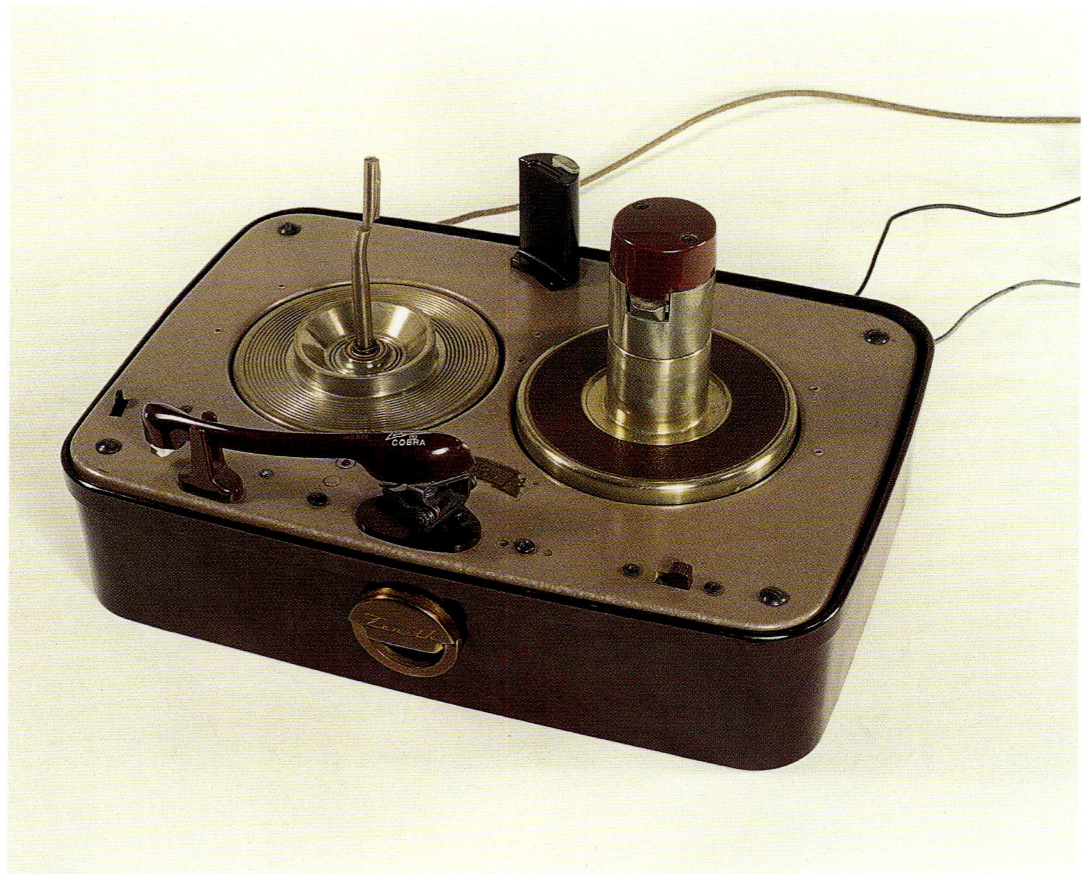

"From a Tops Record Sleeve"

Chapter Five, 1950-1977: Fads, Fashions and Final Frontiers

5-1. This RCA 45rpm player, Model "45-EY-4," with its chestnut plastic cabinet strongly suggests the look of TVs from the late 1940s and early 1950s. A 1950 advertisement from *Look* magazine (for the similar "45-EY-3") stated, "Smart, maroon plastic case… Pile on a whole stack of the non-breakable, 7-inch records – each can play as long as an ordinary 10" or 12", lasts much longer, costs less." *Courtesy of Dan and Sandy Krygier.* (Value code: I)

5-2. The interior of the "45-EY-4" was enlivened by two features: the familiar Nipper trademark inside the lid, and the vivid cap on the thick 45 rpm spindle. The eye was drawn at once to the enlarged spindle, leaving no doubt that Columbia's conventionally shaped 33 1/3 rpm disc (with traditional 1/4" spindle hole) was utterly incompatible. *Courtesy of Dan and Sandy Krygier.*

5-3. Although it very much resembled other RCA 45 players of its vintage (1950) the "45-J-3" (built by Oak Industries for RCA) had one outstanding design feature. The parabolic swoop of the pick-up arm disrupted the angularity of the instrument, like a boomerang suspended in flight. *Courtesy of Bob Havalack "Eager Bros. Music."* (Value code: I)

5-4. This extraordinary phonograph was manufactured by the Ristau brothers (Alfred, Arnold and Harold) in Kaukauna, Wisconsin. Their products were usually coin-operated, but in 1951 they offered this home model. The Ristaucrat "Dial-O-Matic" model "2020" was capable of playing up to 18 45 rpm records in sequence. *Courtesy of Bob Havalack "Eager Bros. Music."* (Value code: H)

5-5. The Ristaucrat mechanism lifted the entire stack and repeated the program for any number of preset auditions. Customized programs could be set by adjusting the dials on the front panel, hence the model's name. Technologically, it was state of the art, but its styling was out of date. *Courtesy of Bob Havalack "Eager Bros. Music."*

5-6. RCA's "Alice in Wonderland" 45 player (Model "45-EY-26") was decorated with scenes from the Walt Disney cartoon of the same name, released in 1951. An advertisement from *Look* magazine that year rhapsodized, "Even the smallest tots can play RCA Victor's Kiddies' '45' phonograph... music at the touch of a button. Gayly [sic] decorated with Walt Disney 'Alice in Wonderland' characters, this complete 'Victrola' 45 phonograph is a perfect gift for young folk... only $29.95." *Courtesy of Dan and Sandy Krygier.* (Value code: H)

5-7. The "6-JY-1C" was one of the 45-rpm players offered by RCA, often designed to be plugged in to an existing radio, phonograph or television. In contrast to the darker colors of many early RCA 45-rpm units, this example was finished in pastel green and white, signaling a switch to the lighthearted colors of the 1950s. *Courtesy of Dave Barker.* (Value code: I)

Right:
5-8. This truly portable RCA radio-45 player operated on batteries. The salmon-colored highlights remind us that plastic was a joyful thing in the 1950s. *Courtesy of Waves.* (Value code: I)

5-9. This Tele-tone 45 player, circa 1953, was a simple, straightforward instrument – yet distinguished by the contrast of its pastel color-scheme with the stark black and white of the turntable components. *Courtesy of Waves.* (Value code: I)

5-10. A 1952 advertisement promoting Zenith instruments, including, "The New Malabar Table Radio-Phonograph – Features the sensational *Cobra-Matic* Record Changer that will play any speed record from 10 to 85 RPM, including 33 1/3, 45 and 78 RPM and the proposed new 16 RPM records." *Courtesy of Greg Bogantz.*

5-11. The RCA "6-JM-1" (1955) was a 45 player in the tradition of the Philco "1201" (see illustration 4-16). The "magic" of "no hands" technology (which RCA called "Slide-O-Matic") played the record automatically. The 45 record was slipped into the slot, and the gold handle was raised, which lifted the record up to the hidden stylus. Such a simple approach had been tried before, and would be tried again. The "Slide-O-Matic" system appeared briefly, and the "old fashioned" pick-up arm continued to reign supreme for another 35 years. *Courtesy of Kurt Wilmarth "Echo-Tone Music."* (Value code: I)

Below:
5-12. The blond finish and splayed spindle legs of this RCA "SHC-9" Stereo are *haut Fifties*. This minimalist hi-fi, with its even more minimal "SHS-12" auxiliary speaker sitting atop, recalls the "shoebox" architecture of the period: identical suburban ranch houses, stunted strip shopping plazas and public buildings lacking visual interest. *Courtesy of Daniel Gustafson.* (Value code: I)

5-13. The interior of the RCA "SHC-9." Although we may criticize this boxy machine for its lack of style, deeper contemplation proves revealing. Psychologists specializing in stress management tell us that complete relaxation is not merely a state of emptiness or absence of jarring thoughts, but the *presence* of elements more difficult to detect. With this in mind, we begin to notice that the knobs on the RCA radio are shaped like little sombreros, the changer arms are well-thought-out scoops and a pyramidal subtext is evident in the "stereo" tag on the pick-up arm, the record-changer support and even the angled legs of the cabinet (see previous illustration). *Courtesy of Daniel Gustafson.*

5-14. Looking as if it might have scuttled out of a flying saucer, this Voice of Music Model "560-A," offered during 1956 to replace the nearly identical "560," exemplified "Atomic Modern" styling. Voice of Music produced many well-built instruments during its long life, and here we see how the company could outdo its competitors at the blond cabinet game. An advertisement for the "560" from the April 1955 issue of *High Fidelity* magazine extolled, "Your choice of either African mahogany or Champagne blonde hand-rubbed finish at the same price…" The simple blond box is rendered almost elegant by the stark, angular base. The asymmetrical placement of the speaker opening and its boldly textured covering lent visual relief. This instrument was available as either a table model, or equipped with "No.7382" legs (brass finish shown) at additional cost. *Courtesy of Gary Stork, The V-M Historical Collection.* (Value code: I)

5-15. The interior of the "560-A" displayed the popular 1950s look of gold components, with controls brightly demarcated in red. As the 1955 advertisement stated, "With Tone-o-matic you accurately control the precise – perfect – balance of reproduction… V-M's Tri-o-matic Spindle gently *lowers* records of all three sizes and speeds [7", 10", 12" and 33 1/3, 45, 78] to the spindle shelf where they are silently air-cushion-dropped to the turntable. After the last record plays, Siesta Switch shuts off even the amplifier." The lucky bachelor owner of a V-M "560" could focus his full attention on his date, and the machine would handle the Johnny Mathis records as reliably as a discreet retainer. *Courtesy of Gary Stork, The V-M Historical Collection.*

5-16. The styling of the 1955 RCA "6HF5" evokes roadside motel architecture of the period. The eye is drawn horizontally across the object, a far cry from the vertical orientation of phonographs during the 'teens and 1920s. More pertinently, the blond finish of this instrument links it unmistakably to mid-1950s fashion. *Courtesy of John A. Reinicke.* (Value code: I)

5-17. The interior of the "6HF5" was merely utilitarian, but the soft salmon of the 45 adapter reminds us that even this simple arrangement could be enlivened by a splash of 1950s pastel. *Courtesy of John A. Reinicke.*

5-18. Voice of Music instruments as contemporary art. A page from the company's 1958 "Fashions in High Fidelity" sales promotion booklet. Included at the lower left is the model shown in illustration 5-14, known at this time as the "562." All that's missing is Audrey Hepburn in a big hat! *Courtesy of Gary Stork, The V-M Historical Collection.*

5-19. When the Guild Radio and Television Company chose a name for this peculiar hi-fi set in the image of an antique, it avoided some of the more egregious misuses of vintage talking machine nomenclature such as "gramma-phone" (presumably something your old gramma once listened to). Instead, the company selected "Grafonola," a genuine term from the 'teens though one unfortunately never applied to external horn instruments. That's a minor point – it could have been much worse. The "Early American" look of the cabinet was a bit jarring, but the overall design and proportions of this machine were in keeping with the subject they imitated (except for the optional legs shown here). For authenticity, a speaker was hidden within the "horn." How curious that the phonograph had gone from imitating other objects to finally imitating itself. Like an octogenarian slipping into a *bustier*, the phonograph was out to recapture its youth. *Courtesy of Doug Anderson.* (Value code: I)

5-20. Original instructions for the Guild "Grafonola." *Courtesy of John P. Andolina, Jr., "The Early Soundman."*

5-21. A tri-fold flyer advertising the Guild "Grafonola." Original price (as quoted in the August 1956 *Billboard*) was $199.95.

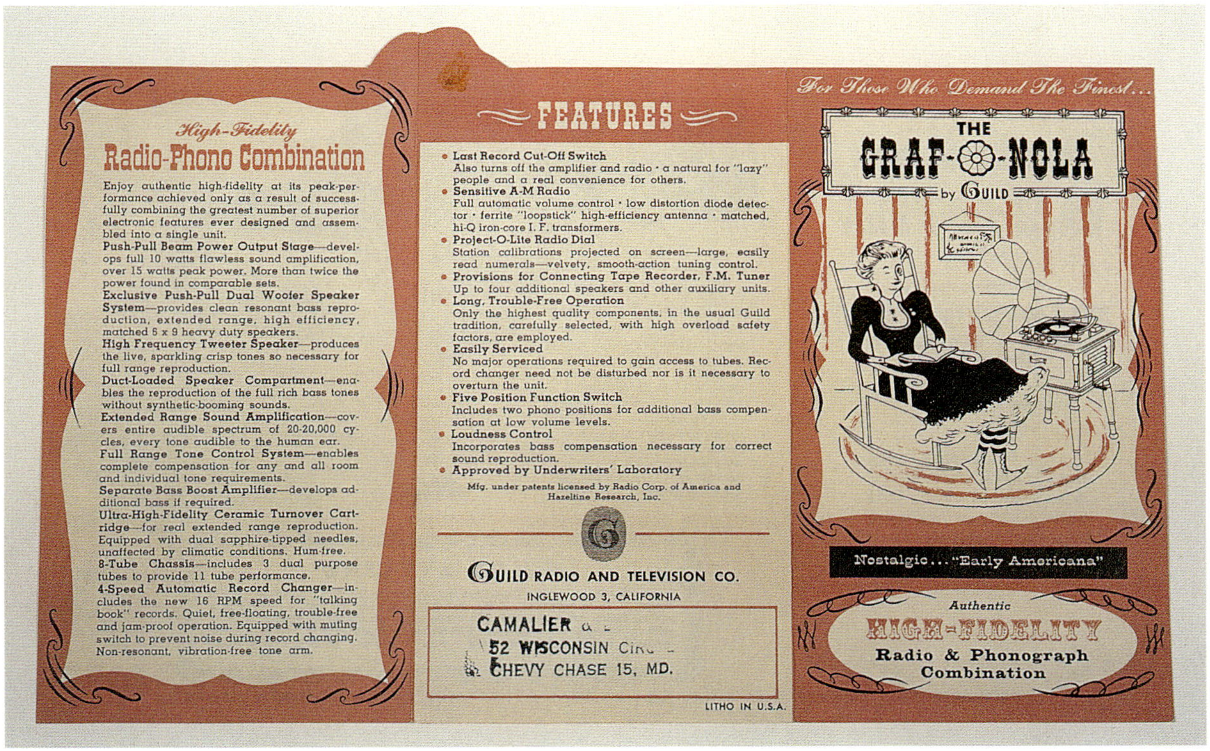

5-22. Reverse of the above tri-fold, illustrated with an "Olde Tyme" flavor. The depiction of the Edwardian lady is no more accurate than the faithfulness of the "Grafonola" to its subject.

5-23. This pudgy Decca "DP910" 45 portable from 1957 displayed the "two-tone" color scheme that was a signature of the 1950s. *Courtesy of Kurt Wilmarth "Echo-Tone Music."* (Value code: H)

5-24. The "DP910" opened like a lady's overnight case (see illustrations 2-41 and 2-59). Note the stark contrast of the technical components. During the late 1950s, black and white record changers became popular. *Courtesy of Kurt Wilmarth "Echo-Tone Music."*

5-25. This 1957 RCA "8-EY-3/HE" looks like something from the school AV cart, except for the pastel two-tone color scheme. This was the last exclusive 45 player offered by RCA for home use. *Courtesy of Kurt Wilmarth "Echo-Tone Music."* (Value code: H)

5-26. This Voice of Music Model "1290" monaural hi-fi, 1957-58, recalled the post-war styling conventions of ten years earlier. Compare this instrument to those pictured in illustrations 4-20 and 4-47. Only the beveled base upon which the body of the cabinet sat gave any hint of lightness to the design. *Courtesy of Dave Barker.* (Value code: I)

Right:
5-27. The Model "1290" at home among the flagstones and fake paneling. A page from Voice of Music's 1958 "Fashions in High Fidelity" booklet. Note how much the appearance of this machine was improved by the blond cabinet. *Courtesy of Gary Stork, The V-M Historical Collection.*

5-28. In the 1950s, pastel shades began to be substituted for the conservative browns and somber tones popular in sound equipment during the previous twenty years. This Voice of Music Model "1260" (1956-59) is typical of inexpensive portable hi-fi sets of the late 1950s and early 1960s: a central lid, frontal speaker and plastic handle for carrying. The salmon of the lid and base buffered the black and white contrast of the interior. *Courtesy of Dave Barker.* (Value code: I)

Right:
5-29. Voice of Music's 1958 "Fashions in High Fidelity" featured the sporty Model "1260" (here illustrated in blue) in a setting as familiar as the pegboard of Dad's basement workroom. *Courtesy of Gary Stork, The V-M Historical Collection.*

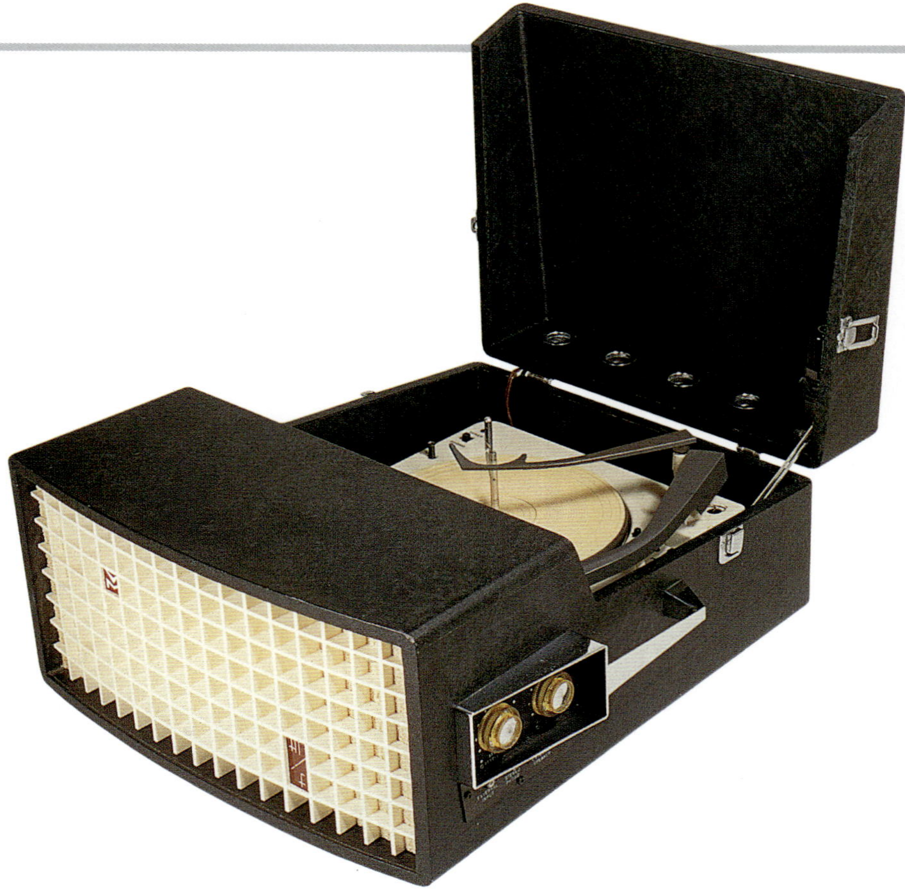

5-30. In 1958, Voice of Music's Model "557" "Ultra-Deluxe" portable stereo (requiring an auxiliary external speaker for stereo) was as sober as a jet intake, yet stylish (with its black and white and chrome) as a prop out of a Fellini movie. *Courtesy of Gary Stork, The V-M Historical Collection.* (Value code: H)

5-31. From Voice of Music's "Fashion's in High Fidelity" of 1958. The elegant contrasts of the Model "557" amid a queasy green layout. *Courtesy of Gary Stork, The V-M Historical Collection.*

5-32. A Voice of Music advertising kite measuring 35" high. *Courtesy of Gary Stork, The V-M Historical Collection.*

5-33. The 1957 Magnavox "TP264B" phonograph sported the decorated faux-leather top surface popular on certain furniture of the time. Like many tabletop instruments of the 1950s, it was also available as a free-standing unit on four slender metal legs. The brown and gold color scheme was interrupted only by the two pieces of chrome trim in the grille. *Courtesy of John A. Reinicke.* (Value code: I)

5-34. The very last 45 player manufactured by RCA was not designed for home use, but to capitalize on America's growing car culture. The "AP-1" was standard equipment on certain 1959-61 Plymouths, where it played through the existing radio. Though it functioned cleverly (a stack of 45s played *upside-down*), the player had a limited future. It was soon discovered that a bumpy road would transform "The 1001 Strings" into "The Fiddling Powers Family." The front of the instrument was hinged to allow access to the turntable. Unit measures 12" wide x 9" deep x 6" high. *Courtesy of Bob Havalack "Eager Bros. Music."* (Value code: G)

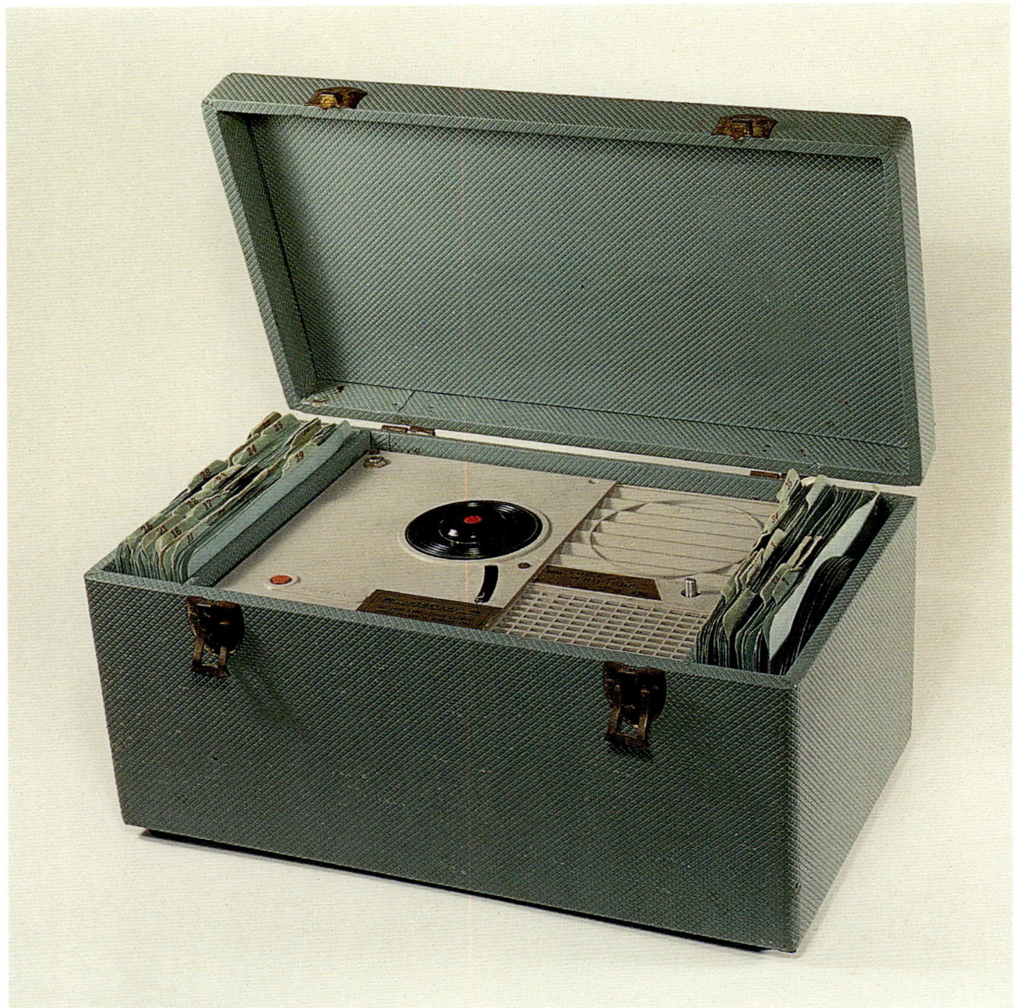

5-35. "The Voice of Scripture," in comforting, quilted seafoam green, was as easy on the eyes as it was on the spirit. The Hawthorne Christian School of Hawthorne, California, offered the entire Bible recorded on 16 rpm (technically 16 2/3 rpm) 7" records. The stylus played the underside of the records. *Courtesy of Bob Havalack "Eager Bros. Music."* (Value code: VR)

5-36. The Voice of Music Model "1000" of 1957-59 had the understated, urbane lines of 1950s "Contemporary" furniture. Made of solid walnut, with a technical complement which included reel-to-reel tape recorder, radio and stereo changer (with Stere-o-matic), the Model "1000" weighed-in at a formidable 214 pounds. *Courtesy of Gary Stork, The V-M Historical Collection.* (Value code: VR)*The V-M Historical Collection.* (Value code: VR)

5-37. The "1000" measured five feet across, and had three separately accessible playing compartments. In 1959, this instrument cost $1000.00, and we seem to recognize in this a return to the practice of matching model number to retail price, favored by phonograph companies during the 1905-1920 period. It was only a coincidence, however, since the Model "1000" had been introduced at $1,125.00. *Courtesy of Gary Stork, The V-M Historical Collection.*

5-38. The striking black control panel in the central compartment recalls battery radio face-plates of the 1920s, then considered a *de rigueur* look for radio equipment. Interestingly, the push buttons popular in 1940s sound equipment also have been revived. *Courtesy of Gary Stork, The V-M Historical Collection*

5-39. A V-M dealer sales sheet for the Model "1000." The "starburst" clock at the upper left was an icon of the American home during the late 1950s and early '60s. *Courtesy of Gary Stork, The V-M Historical Collection.*

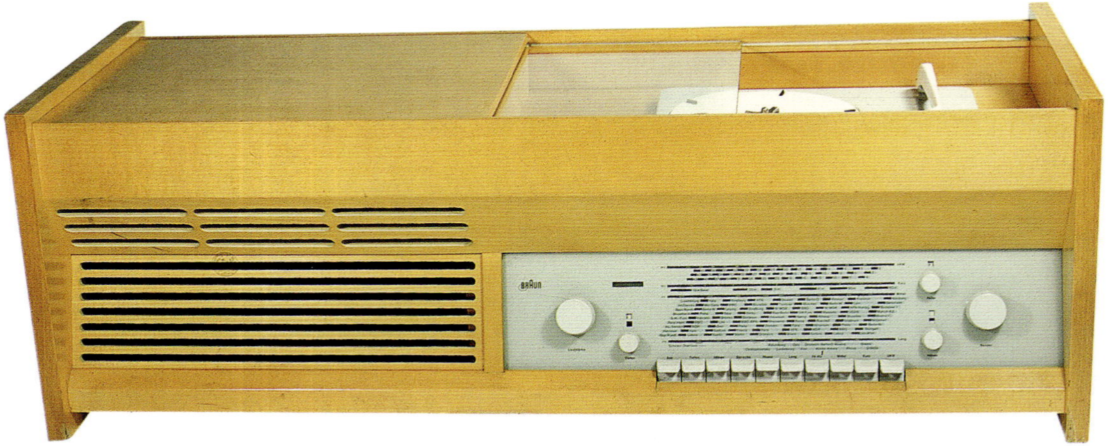

5-40. The Braun "PKG 5-8" AM-FM-short wave radio-phonograph, circa 1960, had the cool, clean look of German precision engineering. The breezy openness of the design celebrated, rather than disguised, the utilitarian nature of the instrument. It could play at 16 (spoken word), 33 1/3, 45 and 78 rpm. *Courtesy of Waves.* (Value code: VR)

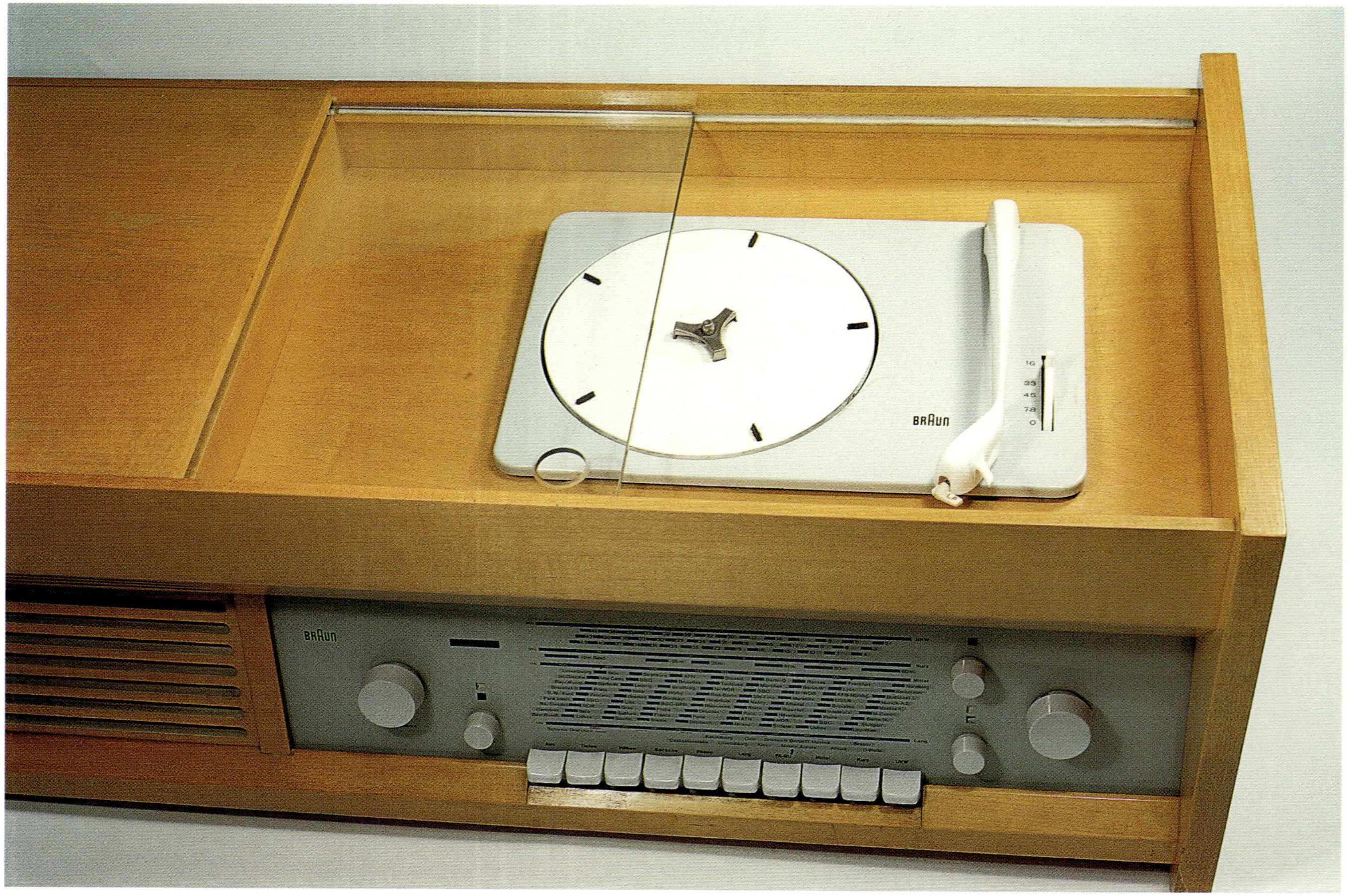

5-41. The phonograph was minimalist, but crisp as a military salute. In 1957, Braun had won the grand prize at the XI Triennale in Milan for the design of its "Phonosuper SK55." The "PKG 5-8" was a direct descendant of the "SK55," which had exemplified rectilinear simplicity. *Courtesy of Waves.*

5-42. The Voice of Music "Songstar" of 1960, a portable stereo phonograph with detachable speakers. The translucent knobs and busy speaker cloth were the most interesting touches. Technically, the "Songstar" was ahead of its time. Its speaker arrangement, featuring a 6"x 9" speaker for lower frequency, and detachable 4" speakers for mid-to-high range is the same plan employed by today's "home theatres." *Courtesy of Gary Stork, the V-M Historical Collection.* (Value code: H)

5-43. On the brink of the Beatles, small-town college girls enjoy echoes of the 1950s on an aging hi-fi from home. From a 1963 college yearbook.

5-44. The RCA "TMP-11" of 1960 had a quasi-colonial look, with its speakers housed obtrusively in the scalloped and cloth-covered façade. *Courtesy of John A. Reinicke.* (Value code: I)

Below:
5-45. The interior of the RCA "TMP-11," however, was definitely "Detroit Modern." The visual emphasis, with control knobs placed evenly amid geometric forms, was borrowed from contemporary automobile dashboard design. Salmon-colored highlights on the 45 adapter and pick-up arm continued the accent shade favored by RCA during the 1950s. *Courtesy of John A. Reinicke.*

5-46. Contemporary 1960s styling with a European flair was evident in this Blaupunkt "New Yorker" of the 1958-1961 period. The high-quality walnut cabinet incorporated rounded corners, prominently displayed speaker cloth, and splayed legs typical of the time. (Value code: I)

Below:
5-47. The upper sliding doors hid an impressive control panel whereby the stereo, AM, FM and short wave bands could be selected. The white plastic buttons would soon give way to a fashion for darker colors, and the large radio dial decorated with the names of European cities would very soon be viewed as archaic. This Blaupunkt "New Yorker" is a virtual "snapshot" of what an early 1960s stereo *should* look like.

5-48. This Magnavox "4ST655A" Stereo, circa 1963, emerged from a time of captain's chairs, corner hutches, plaid wing-backed sofas and eagle-festooned colonial bric-a-brac. The Early American styling of this instrument, in obligatory maple, evokes the living rooms in which so many Americans watched the *Ozzie and Harriet Show*. As the importance of multiple speakers grew, so did the area devoted to speaker cloth. *Courtesy of Greg Bogantz.* (Value code: I)

5-49. A sliding panel revealed the Collaro/Magnavox "Micromatic" changer. Note how the stock mahogany-finish trim of the AM/FM radio conflicted with the uncommonly dark maple of the cabinet. *Courtesy of Greg Bogantz.*

5-50. A 1963 advertisement for the General Electric "Custom Decorator," a stereo with a precision Garrard turntable (to up-grade GE's normally pedestrian audio line), and an innovative approach to housing its speakers. The "speaker-wings" swung outward and could be detached for maximum separation. V-M's "Songstar" had used a simpler version of this design (see illustration 5-42). *Courtesy of Greg Bogantz.*

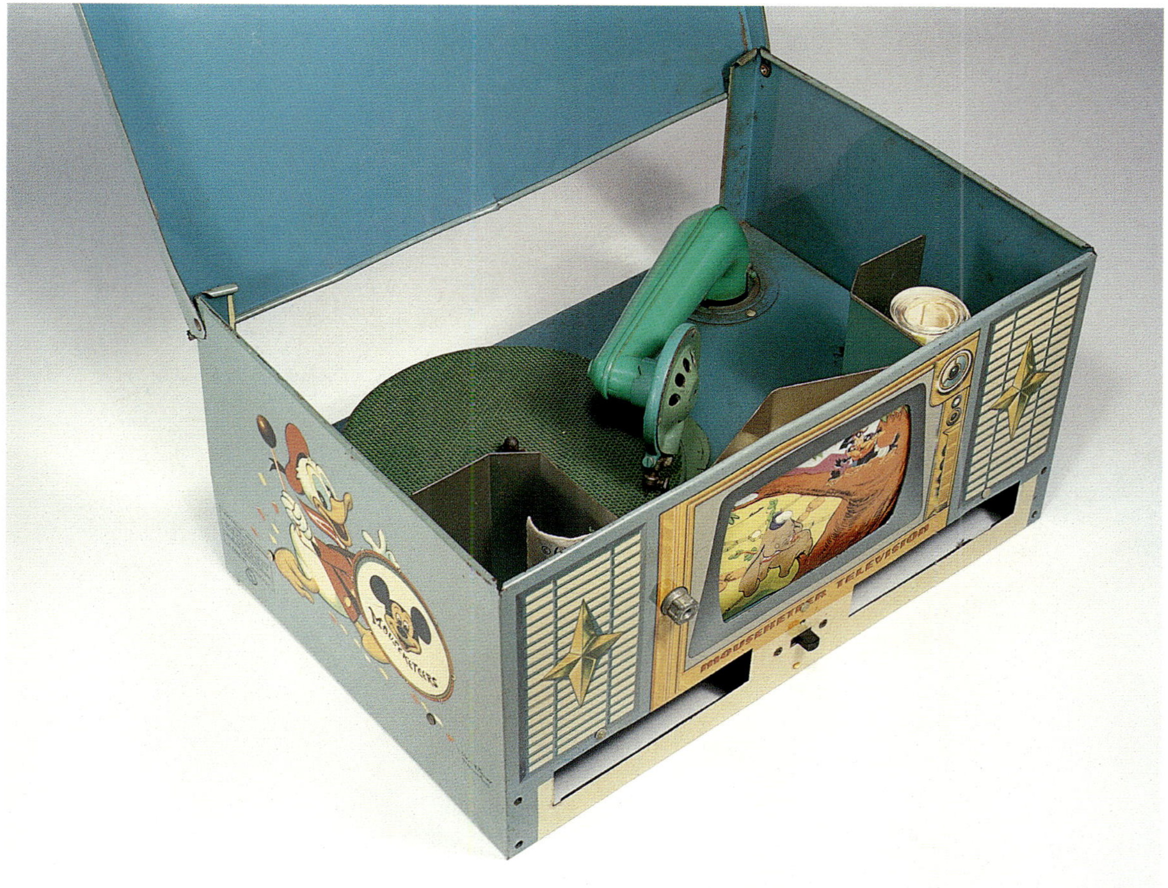

5-51. The acoustic phonograph persisted for decades in children's toys. Witness this "Mouseketeer Television," Model "300-TV" made by Superior, T. Conn Incorporated, Brooklyn, New York. A 78 rpm record was played with a steel needle while the paper roll program scrolled across the "screen." The sound was deflected out the two slots in the lower front, not through the lithographed "speaker" grilles on each side. *Courtesy of Omri Ben-Zion.* (Value code: H)

5-52. For unadulterated kitsch, few phonographs could surpass the Koronette Model "VESUV R10." Manufactured in East Germany in the early 1960s, this radio-phonograph masqueraded as a fireplace, complete with glowing "fire" (via a 100-watt lightbulb), faux marble, and polished walnut veneer over particleboard. Just the thing for a swinging '60s bachelor pad. *Courtesy of the Randy Liebermann collection.* (Value code: VR)

5-53. Opening the two doors allows the partying to begin, komrades! You can almost hear the words, "We ARE two wild and crazy guys!" The swiveling red velour bar at left is illuminated by florescent bulbs and backed by a mirrored surface. And to the right, a feast of broadcast and recorded entertainment awaited.... Behind the pleated red speaker cloth on each side of the fireplace reside one 5"x 8" speaker and two 4" speakers, all made by VEB Elektroakustik of Leipzig, DDR. Could it be that Model "VESUV" referred to the fires of Vesuvius recreated by the fireplace effect, or the Vesuvian passions aroused in some Eastern European James Bond? *Courtesy of the Randy Liebermann collection.*

5-54. The entertainment center consists of a "Phono-Sonic" radio with three shortwave bands, as well as AM and FM Stereo. The turntable is a BSR, made in England. The blond wood and white knobs convey a "Fifties look" within the farcical cabinet. Though this Koronette was marked for export, one assumes that similar entertainment centers were distributed behind the Iron Curtain. *Courtesy of the Randy Liebermann collection.*

5-55. The Columbia Stereo, Model "360," had a look that recalled the front end of a 1962 Dodge – evidence that style is pervasive and driven by community vision. *Courtesy of Waves.* (Value code: H)

5-56. The "360" contained a Garrard turntable, considered high quality equipment at the time. This particular model was made in England by Pye, Limited. *Courtesy of Waves.*

5-57. A Voice of Music glass promotional ashtray, used as an employee favor in the early 1960s. Ashtray measures 3 1/4" x 8 3/4". *Courtesy of Gary Stork, The V-M Historical Collection.*

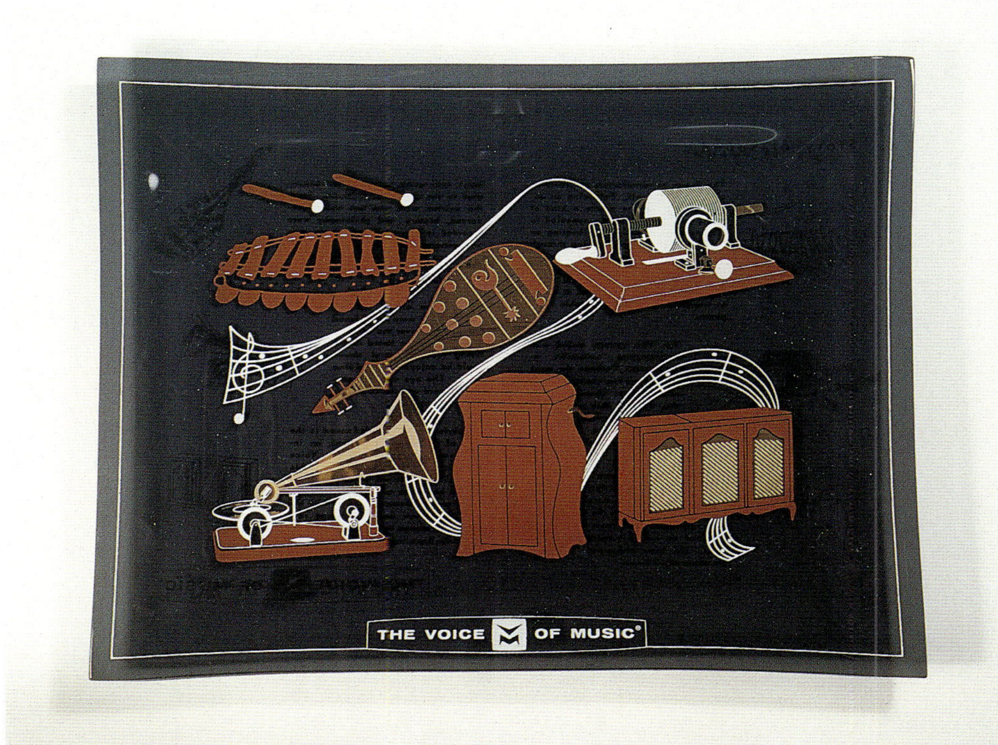

5-58. A Voice of Music glass promotional ashtray with a "history of sound" theme. Considering the rather freeform depiction of the Victrola at the bottom center, it is remarkable that the Edison "Tinfoil" Phonograph (1877) at the top right, and the Berliner hand-driven Gramophone (1895) at the bottom left were so accurately rendered. Ashtray measures 7 1/2" x 9". *Courtesy of Gary Stork, The V-M Historical Collection.*

5-59. The sensory overload of the 1960s included visual as well as musical innovation. Shown is an LP storage box with an "op art" design. The real originator of op art, British contemporary painter Bridget Riley (whose black-and-white canvases from the early 1960s unintentionally began the fad), was largely forgotten as companies rushed to decorate everything from dresses to furniture with geometric designs.

5-60. Whereas the Magnavox shown in illustration 5-48 could not rightly be called a replica of any practical piece of furniture (despite the false pulls on the front panel), this Voice of Music "627" was an attempt to disguise a phonograph as an ersatz antique. This instrument was designed as a table model, but the four-legged stand on which it sits was available at extra cost. *Courtesy of John A. Reinicke.* (Value code: VR)

5-61. Within this Early American desk (expectedly, in maple) was a stereo changer and AM/FM tuner. Note the speaker cloth on the sides. This instrument was sold 1968 - 1971, during a period when the Voice of Music Corporation, which previously had been ubiquitous in the audio market, was attempting to diversify its line. *Courtesy of John A. Reinicke.*

5-62. An original design department drawing of the Model "627" salvaged from The Voice of Music factory in Benton Harbor, Michigan. Note that an ink bottle and a quill pen were original accessories, missing in the example we illustrated. *Courtesy of John A. Reinicke.*

5-63. The owner's manual for the Model "627," one of two Voice of Music "Table Toppers" (so-called because the machines were sold separately from their stands). Note the "Contemporary" model below had already become the featureless box that would define the phonograph in the future. *Courtesy of John A. Reinicke.*

5-64. In 1972, Voice of Music flogged gender bias like candy with its "Plush Pink" (shown here) and several other inexpensive specialty stereos. The flushed complexion of this model (No."338PK") left no doubt of a marketing trajectory toward the feminine. *Courtesy of Greg Bogantz.* (Value code: I)

Below:
5-65. Beginning in the early 1960s, inexpensive portable stereos with this same physical layout had become very popular (General Electric advertised its "Trimline" styling in 1961), and remained in fashion for over ten years. The pivoting central changer was flanked by speakers. The minimal controls had negligible selectivity, the better, perhaps, to concentrate on *heavy* sounds. "Plush Pink" sold for $62.50, a bargain in 1972. *Courtesy of Greg Bogantz.*

5-66. A page from the 1973 Voice of Music catalogue. Quadraphonic systems, featuring four channel sound, once seemed ready to replace ordinary two channel stereo. Quadraphonic eventually proved a rather brief fad in the general market, but this catalogue was issued during its heyday. The two upper models, with garish color-schemes and round speakers covered in "op art" inspired cloth, were aimed at the young and the restless. Below, the future is now: dark, conservative component systems would be the final incarnation of the phonograph. *Courtesy of Gary Stork, The V-M Historical Collection.*

5-67. The good, the bad and the ugly from V-M's 1973 catalogue. "Green Goodie" (Model "346 GN-2") glowed eerily in the middle of the page, like Mom's lime gelatin dessert. Below, "Orange Peel" (Model "346 RG-2") puckered the mind in a citrus fantasy. At the top, "L'il Lavendar" (Model "337 LV") was, "Different – like today's music!" Finally, "Yankee Doodle" (Model "337 WH") anticipated Bicentennial merchandising. This page is a veritable textbook study in "Pop Modern" styling. The V-M Corporation struggled to find a niche in the evolving audio environment of the 1970s, but its home stereo line collapsed in 1975, and the company spiraled into bankruptcy. *Courtesy of Gary Stork, The V-M Historical Collection.*

5-68. The Clairtone "Project G" Model "2000" must have represented a terrible presence in the homes of its owners – even its model name sounds ominous. Certainly, there was no way for this behemoth to blend unobtrusively with existing décor, but beyond that, the Clairtone might have stirred unsettling subconscious images: naval mines, immense bowling balls or the Death Star from *Star Wars*. *Courtesy of Richard Storms.* (Value code: VR)

5-69. The tambour top of the Clairtone (rosewood finish) slid aside to reveal a radio, phonograph turntable and space for storing LPs in garish red velour luxury. *Courtesy of Richard Storms.*

5-70. This instrument was manufactured by the Clairtone Sound Corporation of Canada in the early 1970s. *Courtesy of Richard Storms.*

5-71. The mesh Death Star speakers which flanked the console body of the machine created an unfathomably "polarized" look. *Courtesy of Richard Storms.*

5-72. This Morse "Electrophonic" was more of an "entertainment caddy" than a stereo set. When the movie *The Sting* revived interest in the "good old" days, the mid-1970s went awash in fake-oak Formica (supplementing the formerly obligatory fake-walnut). The clock was a curious touch, but gone was the radio which had been the phonograph's strange bedfellow for nearly 50 years. *Courtesy of Doug Anderson.* (Value code: I)

5-73. Backyard barbecue grill or radio-phonograph-eight track combination? This 1970s Brother "Primus II," is an excellent example of the "pedestal" look that enjoyed brief popularity during "Disco Modern." Courtesy of Waves. (Value code: H)

5-74. With domed plastic cover removed, the "Primus II" displays cockpit styling suitable for any intergalactic cruiser. The attending speakers are the least inspiring part of the package. *Courtesy of Waves.*

5-75. This Electrohome (of Canada) phonograph-radio, from the early 1970s, presaged the "Star Wars" look. In fact, without external speakers visible it might be a droid rather than a record player. The clever manner in which the smoked plastic dome rested in the open position showed an inventive concern for the consumer. Courtesy of Waves. (Value code: H)

5-76. This illustration, showing the Electrohome and its two external speakers, is a geometric study in curvilinear forms. The designer has supported the recurring shape with unerring fidelity, or might we say *high fidelity? Courtesy of Waves.*

5-77. For over-the-top luxury, this stereo "listening pod" could be described as "disco infernal!" A glitteringly bechained victim of Saturday night fever and his date would snuggle in garish velour splendor while multiple speakers bombarded them with Bee Gees hits. Matching plastic footstools were the ultimate touch, but please remove your platform shoes. As befit a "Disco Modern" audio component, not a speck of wood was used in the construction of this remarkable artifact of the 1970s. *Courtesy of Richard Storms.* (Value code: VR)

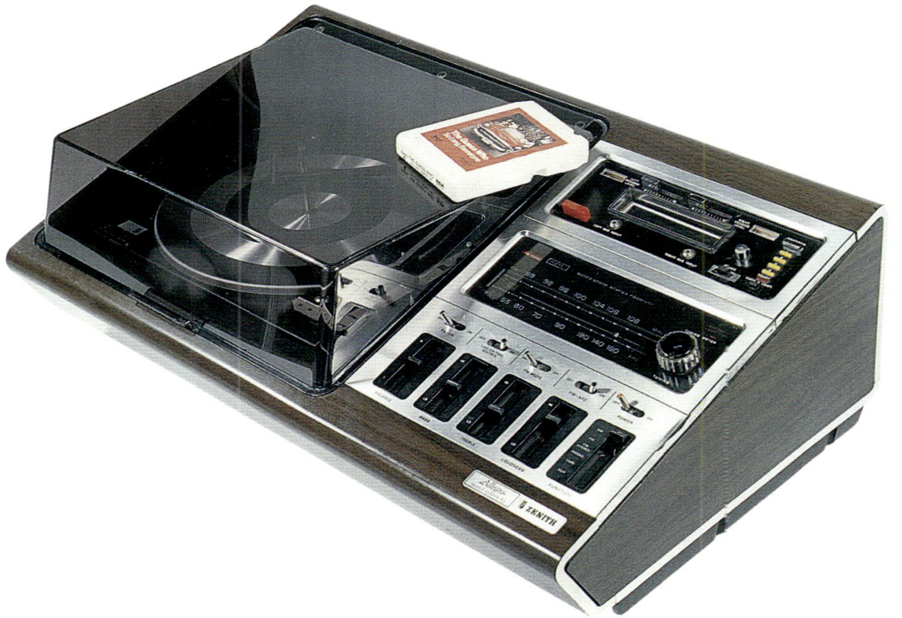

5-78. In 1977, the Zenith "HR596" "Allegro" still conveyed a design statement in an industry where audio systems with featureless, box-like components had become the norm. This elaborate control console might have been extracted from the *Millennium Falcon*: rife with switches, slides and calibrations. In the century since the inception of recorded sound, the phonograph had often masqueraded as furniture or paraded as fashion. After 1984, during its remaining years before being superceded by the CD player, the phonograph was to slip quietly into anonymity, and become just another discreet resident of the "entertainment center." *Courtesy of John A. Reinicke.* (Value code: I)

Bibliography

Association for Recorded Sound Collections, various ed.s: *ARSC Journal*, various issues to present.

Barnum, Frederick O. III: *His Master's Voice in America*, Camden, General Electric Company, 1991.

Barr, Steven C.: *The Almost Complete 78 RPM Record Dating Guide (II)*, Huntington Beach, Yesterday Once Again, 1992.

Baumbach, Robert W.: *Columbia Phonograph Companion, Vol. II (Disc Graphophones and the Grafonola)*, Woodland Hills, Stationery X-Press, 1996.

Baumbach, Robert W.: *Look for the Dog*, Woodland Hills, Stationery X-Press, 1996.

Baumbach, Robert W.: *Victor Data Book*, Los Angeles, Mulholland Press, 2000.

Bayly, E., ed.: *The EMI Collection*, Bournemouth (GB), Talking Machine Review, 1977.

Bayly, E., ed.: *The Talking Machine Review International*, Bournemouth (GB), various issues beginning 1970.

Bill, Edward Lyman, ed.: *The Talking Machine World*, New York, various issues 1905-1929.

Brooks, Tim, ed.: *Little Wonder Records, a History and Discography*, St. Johnsbury (VT), New Amberola Phonograph Company, 2000.

Bryan, Martin, ed.: *The New Amberola Graphic*, St. Johnsbury (VT), various issues 1973 to present.

Charosh, Paul, compiled by: *Berliner Gramophone Records, American Issues 1892-1900*, Westport, Greenwood Press, 1995.

City of London Phonograph and Gramophone Society, various ed.s: *The Hillandale News*, various issues 1960s to present.

Conot, Robert: *A Streak of Luck*, New York, Seaview Books, 1979.

Dethlefson, Ronald, ed.: *Edison Blue Amberol Recordings, 1912-1914*, Brooklyn, APM Press, 1980.

Dethlefson, Ronald, ed.: *Edison Blue Amberol Recordings, 1912-1914* (paper cover), Woodland Hills (CA), Stationery X-Press, 1997.

Dethlefson, Ronald, ed.: *Edison Blue Amberol Recordings, 1915-1929*, Brooklyn, APM Press, 1981.

Dethlefson, Ronald, ed.: *Edison Blue Amberol Recordings, 1915-1929*, second edition, Los Angeles, Mulholland Press, 1999.

Dethlefson, Ronald and Copeland, George: *Pathé Records and Phonographs in America, 1914-1922*, Los Angeles, Mulholland Press, 2000.

Edge, Ruth and Petts, Leonard: *The Collector's Guide to 'His Master's Voice' Nipper Souvenirs*, London, EMI Group, 1997.

Fabrizio, Timothy C. and Paul, George F.: *Antique Phonograph Gadgets, Gizmos and Gimmicks*, Atglen (PA), Schiffer Publishing, Ltd., 1999.

Fabrizio, Timothy C. and Paul, George F.: *Discovering Antique Phonographs*, Atglen (PA), Schiffer Publishing, Ltd., 2000.

Fabrizio, Timothy C. and Paul, George F.: *The Talking Machine, an Illustrated Compendium 1877-1929*, Atglen (PA), Schiffer Publishing, Ltd., 1997.

Fagan, Ted, compiled by: *The Encyclopedic Discography of Victor Recordings, Pre-Matrix Series*, Westport, Greenwood Press, 1983.

Fagan, Ted, compiled by: *The Encyclopedic Discography of Victor Recordings*, Westport, Greenwood Press, 1986.

Frow, George L.: *Edison Cylinder Phonograph Companion*, Woodland Hills, Stationery X-Press, 1994.

Frow, George L.: *The Edison Disc Phonographs and the Diamond Discs*, Sevenoaks, Kent (GB), George L. Frow, 1982.

Gaisberg, Frederick W.: *The Music Goes Round*, North Stratford (NH), Ayer Company Publishers, Inc., 1977 (1942 reprint).

Gelatt, Roland: *The Fabulous Phonograph, revised*, New York, Appleton-Century, 1965 (three ed.s 1955-1977).

Gracyk, Tim, with Hoffmann, Frank: *Popular American Recording Pioneers, 1895-1925*, Binghamton (NY), Haworth Press, 2000.

Gracyk, Tim, ed.: *Victrola and 78 Journal*, Roseville (CA), various issues 1994-1998.

Greenburg, Cara: *Mid-Century Modern*, New York, Harmony Books, (softcover) 1995.

Hardy, William, Adams, Steven and Van de Lemme, Arie: *Encyclopedia of Decorative Styles, 1850-1955*, Secaucus, Wellfleet Press, 1988.

Hatcher, Danny, ed.: *Proceedings of the 1890 Convention of Local Phonograph Companies*, Nashville, Country Music Foundation Press, reprint, 1974.

Hazelcorn, Howard: *A Collector's Guide to the Columbia Spring-Wound Cylinder Graphophone, 1894-1910*, Brooklyn, APM Press, 1976.

Hazelcorn, Howard: *Columbia Phonograph Companion, Volume I: Hazelcorn's Guide to the Columbia Cylinder Graphophone*, Los Angeles, Mulholland Press, 1999.

Hine, Thomas: *Populuxe*, New York, Alfred A. Knopf, 1986.

Hunting, Russell, ed.: *The Phonoscope*, New York, Phonoscope Publishing Company, various issues 1896-1900.

Johnson, E. R. Fenimore: *His Master's Voice Was Eldridge R. Johnson,* Milford (DE), State Media, Inc., 1975.

Koenigsberg, Allen, ed.: *The Antique Phonograph Monthly*, Brooklyn, APM Press, various issues 1972-1993.

Koenigsberg, Allen: *Edison Cylinder Records, 1889-1912, 2nd ed.*, Brooklyn, APM Press, 1988.

Koenigsberg, Allen: *The Patent History of the Phonograph, 1877-1912*, Brooklyn, APM Press, 1991.

Lorenz, Kenneth M.: *Two-Minute... Cylinders of the Columbia Phonograph Company*, Wilmington (DE), Kastlemusick, Inc., 1981.

Maken, Neil: *Hand-Cranked Phonographs*, Huntington Beach, Promar Publishing, 1995.

Marco, Guy A. and Andrews, Frank, ed.s: *Encyclopedia of Recorded Sound in the United States*, New York, Garland Publishing, Inc., 1993.

Martland, Peter: *Since Records Began: EMI: the First Hundred Years*, (distributed by Timber Press, Portland, OR), Amadeus Press, 1997.

Marty, Daniel: *Histoire Illustrée du Phonographe (Illustrated History of the Phonograph)*, Lausanne/Paris, Edita/Lazarus, 1979, (reprinted in various English language editions with slight modifications of the title).

Michigan Antique Phonograph Society, various ed.s.: *In the Groove,* various issues to present.

Moore, Jerrold Northrop: *A Matter of Records*, New York, Taplinger Publishing Company, 1977.

Moore, Wendell, ed.: *The Edison Phonograph Monthly*, various anthologies 1903-1916, New Albany (IN), Wendell Moore Publications.

Proudfoot, Christopher: *Collecting Phonographs and Gramophones*, New York, Mayflower Books, 1980.

Read, Oliver and Welch, Walter L.: *From Tinfoil to Stereo*, Indianapolis, Howard W. Sams and Company, Inc., 1959.

Reiss, Eric L.: *The Compleat Talking Machine*, Chandler (AZ), Sonoran Publishing, (revised)1998.

Rust, Brian, compiled by: *Discography of Historical Records on Cylinders and 78s*, Westport, Greenwood Press, 1979.

Sherman, Michael W. (with Moran, William R. and Nauck, Kurt R. III): *Collector's Guide to Victor Records*, Dallas, Monarch Record Enterprises, 1992.

Sherman, Michael W. and Nauck, Kurt R. III: *Note the Notes, An Illustrated History of the Columbia 78 rpm Record Label, 1901-1958*, Monarch Record Enterprises, New Orleans, 1998.

Sutton, Allan: *Directory of American Disc Record Brands and Manufacturers, 1891-1943*, Westport, Greenwood Press, 1994.

Sutton, Allan and Nauck, Kurt: *American Record Labels and Companies, an Encyclopedia (1891-1943)* [with CD-ROM], Denver, Mainspring Press, 2000.

Welch, Walter L. and Burt, Leah Brodbeck Stenzel: *From Tinfoil to Stereo, 1877-1929*, Gainesville, University Press of Florida, 1994.

Wile, Raymond R.and Dethlefson, Ronald: *Edison Diamond Disc Recreations, Records and Artists 1910-1929,* Brooklyn, APM Press, 1985.

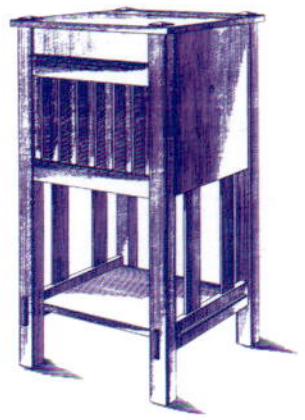

Glossary

AMBEROL: The first four-minute cylinder developed by Edison, 1908-1912. These cylinders are black in color, made of especially hard metallic soap, and were usually sold in green and white containers.

AMBEROLA: The name denoting Edison's line of internal-horn cylinder Phonographs, 1909-1929.

AMPLIFIER: The electronic device which accepts impulses from the PICK-UP, and boosts the signal to a level capable of driving the SPEAKER(S).

AUDIO: Generally used as a prefix to denote sound-reproducing apparatus, as in "AUDIO system," "AUDIO equipment" and "AUDIO components."

BACK MOUNT: A term used to describe a talking machine which uses a back bracket which supports the horn and thus removes its inertia from the SOUNDBOX.

BEDPLATE: The plate (usually metal) to which the upper mechanism of a cylinder talking machine is mounted.

BLANK: A smooth, grooveless unrecorded cylinder.

BLUE AMBEROL: The blue celluloid four-minute cylinder sold by Edison, 1912-1929. The containers for these cylinders were blue until 1917, orange and blue thereafter.

CARRIAGE: The assembly which "carries" the REPRODUCER of a cylinder talking machine across the recording.

CARRIER ARM: Edison nomenclature for the CARRIAGE. The HALF-NUT and its flat supporting spring are attached directly to the carrier arm on nearly all Edison machines.

CHANGER: see "RECORD CHANGER."

COMBINATION ATTACHMENT: Devices offered by Edison to convert pre-1908 Phonographs to play four-minute cylinders in addition to the two-minute variety.

COMPONENT: Rightly, a specific part of a single machine or electronic device, but frequently used to denote a particular link in a chain of sound-reproducing devices, such as the RECORD CHANGER, AMPLIFIER or SPEAKERS.

CONCERT: Edison's trade name for the five-inch diameter cylinder and the Edison Phonograph designed to play it.

CRANE: The support used to mount horns of 15-inch length or greater to cylinder talking machines.

CYLINDER: A geometric form on which entertainment recordings were made, 1877-1929. The earliest sheets of tinfoil were followed by self-supporting cylinders of ozocerite-covered cardboard, stearic acid/paraffin, hard metallic soap, and celluloid. Advantages of the cylinder format included a constant surface speed and the ability to make home recordings.

DIAMOND DISC: Edison's line of disc records and machines, 1912-1929. The Diamond Discs were vertically recorded, and the Diamond Disc Phonographs were one of the very few disc talking machines to use a FEEDSCREW.

DIAPHRAGM: The flexible, circular vibrating membrane of a SOUNDBOX or REPRODUCER which converts mechanical energy to acoustic energy or sound waves. Mica, glass, compressed paper, copper and aluminum were commonly used as diaphragms in early talking machines.

ELBOW: The connection between the horn and the SOUNDBOX or TONE ARM of a disc talking machine. Earliest elbows are made of leather, gradually giving way to metal elbows after 1900.

ENDGATE: A swinging arm which supports one end of the MANDREL on some cylinder talking machines. To mount or remove a cylinder it is therefore necessary to open and close the endgate.

FEEDSCREW: A threaded rod which usually drives a half-nut fixed to the carriage of a cylinder talking machine, thus guiding the reproducer across the record grooves. Certain disc machines used FEEDSCREWS to drive the SOUNDBOX across the record or to move the turntable beneath a fixed soundbox. Similarly, some cylinder machines used feedscrews to drive a mandrel longitudinally beneath a fixed reproducer.

FRONT MOUNT: A term used to describe a disc machine where the horn attaches directly to the SOUNDBOX, and the support arm (or mount) runs directly below and parallel to the HORN. In such an arrangement the support arm points in the same general direction as the bell of the horn.

GRAMOPHONE: Originally the name by which Emile Berliner's disc talking machine was known, it became a generic term to denote any disc playing talking machine, but fell out of use in the United States.

GRAND: Columbia's trade name for the five-inch diameter cylinder and the various Columbia machines designed to play it.

GOVERNOR: The mechanical assembly in a talking machine motor which regulates the speed, usually by limiting the outward movement of spinning weights.

HALF-NUT: An internally threaded metal piece usually in the form of a nut which has been cut in half. The threads of the half-nut correspond to the threads of the FEEDSCREW. As the feedscrew revolves, the half-nut will be propelled along it, thus driving the CARRIAGE of a talking machine.

HI-FI: "High Fidelity" was a descriptive term generally applied to sound reproducing apparatus after the introduction of Long Playing records in 1948, indicating a qualitative value of the reproduction. In its shortened form, it came to represent the sound reproducing unit itself.

LATERAL RECORDING: The type of disc talking machine recording where the information is encoded in sides of the groove. (see VERTICAL RECORDING)

MANDREL: The tapered drum upon which the cylinder record is placed for playing.

MONAURAL: Sound reproduction comprised of one signal, or one channel. In a multiple-speaker STEREO system, a MONAURAL recording will result in the speakers delivering identical signals.

NEEDLE: The point (usually steel) of the SOUNDBOX which rides the grooves of the recording and transmits vibrations to the DIAPHRAGM.

NEEDLE HEAD: A colloquial term for the SOUNDBOX or PICK-UP which accepts vibrations directly from the record.

NEEDLE TIN: A container to store and dispense steel talking machine needles.

NEEDLE SHARPENER (OR CUTTER): A device for repointing disc talking machine needles of fiber, thorn or steel.

PHONOGRAPH: Originally, the name given by Edison to the sound recording/reproducing device he invented in 1877. In American usage, the term has been applied to sound reproducing equipment in general, regardless of age or type of record played.

PICK-UP: The device that converts mechanical vibrations of the NEEDLE or STYLUS into electrical impulses, to be fed through the AMPLIFIER and converted into sound waves by the SPEAKER. The PICK-UP ARM is the levered arm that supports the PICK-UP.

QUADRAPHONIC: Briefly popular during the 1970s, Quadraphonic systems required specially-recorded discs. The reproduced sound was separated into four channels, allowing effects unobtainable with two-channel STEREO.

RECORD CHANGER: That mechanical apparatus in an electrically-operated disc phonograph which automatically changes the record or flips the record to its reverse side. In electric phonographs, the entire mechanical playing mechanism, including the PICK-UP ARM, is sometimes referred to as the CHANGER.

RECORDER: A device comprising a DIAPHRAGM and cutter stylus for recording cylinder or disc records.

REPEATING ATTACHMENT: A device for cylinder or disc talking machines which returns the NEEDLE or STYLUS to the beginning of the record after the selection has played.

REPRODUCER: The component comprising STYLUS, linkage and DIAPHRAGM, which reproduces the sound from the record grooves. This term is most frequently applied to cylinder machines. (See SOUNDBOX)

SHAVER: On a cylinder talking machine, a device which holds a knife in position to pare the recorded surface from a wax cylinder.

SOUND SYSTEM: An inter-related group of COMPONENTS which comprise a complete electric reproduction system.

SOUNDBOX: The component comprising NEEDLE, linkage and DIAPHRAGM, which reproduces the sound from the record grooves. This term is most frequently applied to disc machines. (See REPRODUCER)

SPEAKER: A device comprised of a transducer which converts electrical impulses into sound, usually employing a paper diaphragm.

STEREO: Offered after 1958, stereo (which required specially-recorded discs or selective broadcasts) provided two separate channels or signals, creating more lifelike reproduction. After the introduction of stereo, older, single-channel sound systems were described as MONAURAL.

STYLUS: The point (usually sapphire or diamond) of the REPRODUCER which rides the grooves of the recording and transmits vibrations to the DIAPHRAGM or PICK-UP.

TALKING MACHINE: A general term for sound reproducing devices playing either cylinder or disc records, made prior to the advent of electric reproduction. Specific brands such as Graphophone, Victrola or Amberola all fall within the general category of talking machines. Modern American usage substitutes PHONOGRAPH for the older term "talking machine." For the purposes of historical accuracy, both terms appear frequently in this book.

TONE ARM: A movable hollow tube which conducts sound to the HORN from the SOUNDBOX.

TRUNNION: Rightly, the sleeve visible on either side of the CARRIAGE which slides over the FEEDSCREW shaft on a Graphophone, but commonly used to denote the entire carriage.

TURNTABLE: The round platter which supports a disc record and revolves while the record is being played. After 1960, that separate component in a SOUND SYSTEM which comprised the TURNTABLE and RECORD CHANGER was often referred to simply as the TURNTABLE.

VERTICAL RECORDING: The type of cylinder or disc recording in which the sound vibrations are encoded in the bottom of the groove. (see LATERAL RECORDING)

WAX: Although the material from which many cylinder records were made is generally referred to as "wax," it was more rightly various formulations of metallic soap. The resemblance of this material to conventional wax promotes the common descriptive term.

Value Code Key

Chacun à son goût.

Preparing a value system for a such a diverse number of phonographs, from such a broad spectrum of time periods, is a difficult task. Additionally, some of the items discussed in this book, primarily post WWII hi-fis and stereos, are struggling to find a place in the world of collecting. We believe that the comparatively recent artifacts of audio history deserve as much consideration as older and more venerable pieces. One of the purposes of this book is to draw attention to the areas of audio collecting where considerable growth may be anticipated.

Of the systems commonly in use, we have adopted the "Letter Designated Grading Guide." Needless to say, everyone has his own special axe to grind when it comes to price guides. Since the publication of our book *The Talking Machine, an Illustrated Compendium, 1877-1929*, we have been treated to every sort of theory. Some demanded that the prices be in specific dollar amounts and higher than retail. Some claimed values should be much lower than retail. Others insisted upon a range. No one will be completely satisfied, and some could be agitated by our choice of the letter system. We stand behind our decision. In keeping with precedent, we will designate prices *only* for phonographs, not peripheral items. Items that are one-of-a-kind or uncommonly sold will be rated "very rare" (VR). Please note that "VR" does *not* indicate a price in excess of the scale below. It is merely an indication of infrequently exchanged items.

Needless to say, the people who buy external horn phonographs are frequently not the same people who collect hi-fi sets, and vice versa. Using the tools and knowledge at our disposal (including current Internet auction prices), we have attempted to accurately reflect the current values of all the machines in this book.

It is especially important in the case of electric sound equipment that an item be functional. Non-functional, incomplete and "as is" units can sell for almost nothing. For the purposes of the values in this book, we are referring to functional or restored pieces.

Values will be expressed *in the individual captions* by the following letter codes:

A	\$12,000. and above
B	\$8,000. to \$12,000.
C	\$5,000. to \$8,000.
D	\$3,000. to \$5,000.
E	\$2,000. to \$3,000.
F	\$1,000. to \$2,000.
G	\$500. to \$1000.
H	\$250 to \$500.
I	Under \$250.

In using the value codes, we advise *conservatism*. If you are trying to establish the value of an object in your possession, the natural instinct is to seize upon the highest price in the given range, regardless of condition. This can be a mistake. The condition of the items in the book is nearly always excellent, but the condition of *your* item is the important issue. Condition can affect the value by up to 50% - and occasionally more. The bottom end of a price range has as much validity as the top, though it seems a less attractive prospect. Different examples of the same type of object may represent the entire range of a certain price category, due to condition. The ranges are also intended to be *comparative*. Certain objects in category "H," for instance, may *never* reach \$500.00 no matter how perfectly preserved they are.

Beyond condition, many, many factors affect the price of an "antique." Exact worth is something that can not be fixed precisely in time. We can only hint at it. Our guide is intended to help you understand the value of the objects illustrated, but we vigorously discourage the use of the phrase, "It *books* for such and such a price." True worth is what the market will bear. Quoting books, auction results or what a shop is asking will probably not elevate the price higher than the general market will currently support.

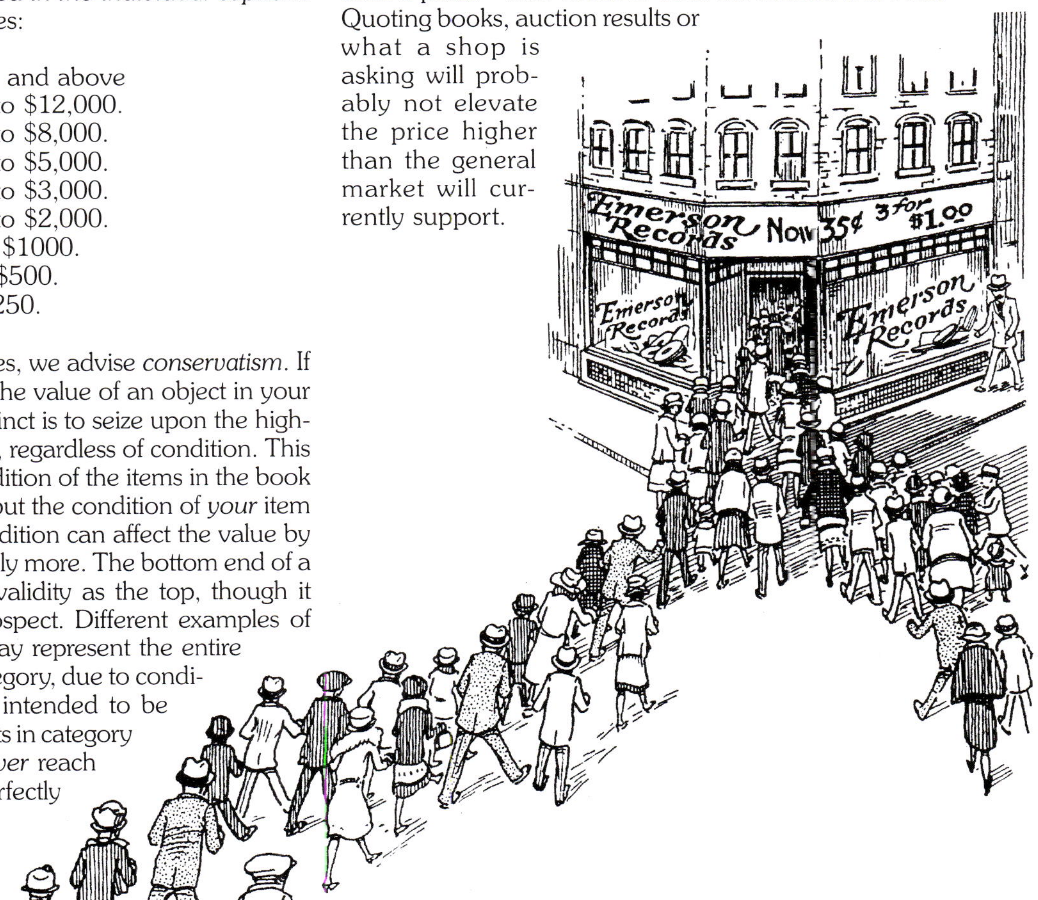

Appendix

The following article on 45 players was written by Kurt Wilmarth and Bob Havalack, with the assistance of the authors of this book.

Madame X

On January 10, 1949, RCA announced to the public the details of its top secret project called "Madame X." The company had designed a totally new system of playing and changing records: the 45 rpm system. Details of the system were issued to dealers on February 19, 1949, and players were made available to the public by April 1, 1949. The new changers incorporated a 1 1/2-inch spindle with an internal mechanism that could change records in 1/45 of a revolution. A new lightweight arm using a crystal cartridge could track a record at 1/16 of an ounce saving wear and tear on the record grooves. The new records were 7-inch diameter.

The first two changers, the "9-EY-3" and "9JY," were housed in maroon Bakelite cabinets. The "9-EY-3" was a complete self contained unit with changer, amplifier, and speaker, while the "9JY" was a changing attachment that could be connected to an amplifier, radio or television to amplify the player's crystal cartridge. The new players had several unique features. One was that the records stacked on the center spindle were rotating at 45 rpm at the same time as the record being played, so that when the next record dropped to the play position it was already revolving at the proper speed. Thus, there was no hesitation in coming up to the required speed.

The record's innovative design incorporated a raised area between the large center hole and the edge of the label, making air spaces between the playing surfaces of the stacked records. The problem of records being scuffed while stacked and dropped was solved by this design element. The new records, made of "Vinylite," came in a variety of colors to ease consumers into locating their favorite type of music. Green vinyl denoted Country, cerise for R&B, yellow records for children, red for classical, and finally blue for international selections.

To compete with Columbia's 7-inch 33 1/3 microgroove records, RCA issued the EP or extended play microgroove 45 record, with additional playing time. From 1949 until early 1958 various players were produced with Bakelite, wood, styrene (two-tone), and metal cabinetry. Some were combined with radios, televisions or full sized consoles. Some were battery operated. A few were produced without the changer mechanism for single play use. By early 1958, all home tabletop stand-alone units were phased out. The final RCA 45 player, the "AP-1" produced from 1959 through 1961, was an inverted changer for use in an automobile only. In 1987, it was announced that, due to sagging sales, the 45 rpm record would no longer be manufactured for the mass market. They are, however, still produced by small independent record companies.

Index

$84.30 for this new

including the "M 10" series of Columbia Double-Disc Records and Record Album

Until July 1, delivered ON APPROVAL by 7500 dealers, payment terms $7 a month—no interest, no extras—and subject to a guarantee of full satisfaction.

Or $75 flat for the instrument alone. No other upright instrument has ever been offered for less than $100, with NO records.

THIS new Grafonola "Princess" is a beautiful mahogany table, ready at all times to serve all the purposes of such a table, worthy of its place in any part of any room in the house, yet likewise ready to provide music of any class you happen to desire, and capable of all the quality of tone of any instrument by which you may happen to judge it.

This newest musical instrument, with the "M 10" outfit of records—a total of 24 different selections, sufficient for an entire evening's delightful entertainment, together with an album to contain them—costs you just $84.30, and until July 1st, may be paid for at the rate of $7 a month, and will be delivered on approval by your dealer. Besides the "M 10" series of records, do not miss the opportunity of hearing this most versatile of all musical instruments render selections by *Lillian Nordica, Olive Fremstad, Mary Garden, Emmy Destinn, Alice Nielsen, Zenatello, David Bispham, Maria Gay* and any of the other great artists whose voices appeal to you.

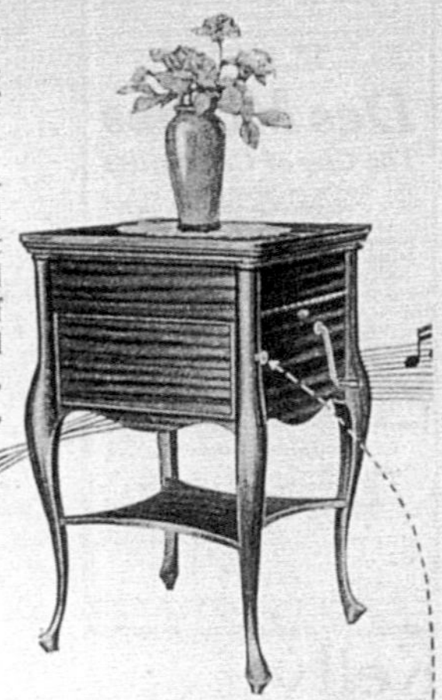

With lid and tone-shutters closed—a completely concealed musical instrument, serving as a handsome small table.

The tone-shutters are opened or closed by turning this little knob.

Go to Your Nearest Dealer

The nearest one of 7500 dealers will place this complete outfit in your home on free trial, giving you an opportunity to combine an evening's entertainment with a thorough demonstration, and guaranteeing the outfit to be exactly as advertised or money refunded.

If by any chance there is no dealer in your neighborhood who is prepared to supply you, drop us a line and we will see that your inquiry is attended to.

Columbia Phonograph Company, General

Box 328, Tribune Building, New York

TORONTO—McKinnon Building. (Prices in Canada plus duty.)

Dealers Everywhere, but the fast-increasing Columbia demand affords splendid opportunities for ***MORE DEALERS.*** Write in for trade particulars. Exclusive selling rights granted where we are not actively represented.

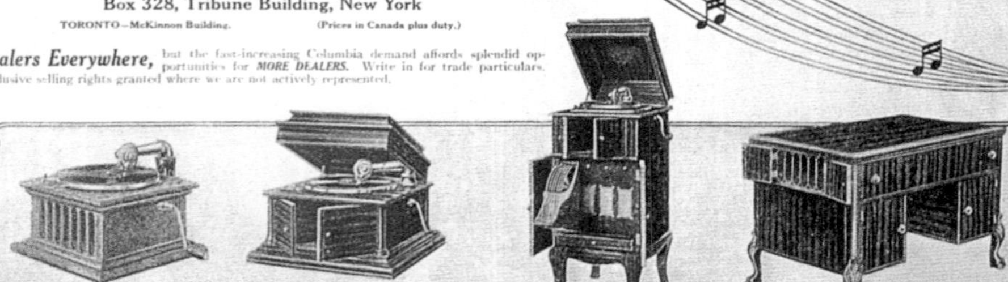

Columbia Hornless Graphophone "Lyric"—$25

Columbia Grafonola "Favorite"—$50

Columbia Grafonola "Nonpareil"—$150

Columbia Grafonola "Regent"—$200

Columbia Grafonola

The "Princess"

First announcement of the *tone-shutters*, replacing the two small doors, and providing more sightly, more effective and more convenient control of tone-volume.

By turning the small knob, it is easy to partly or completely close the tone-shutters at the opening of the sound-chamber, regulating the *volume* of music without affecting its tonal integrity.

An Exclusive Columbia Feature

There are other Columbia Grafonolas at other prices (Graphophones from $17.50 up)